NUREG-1650, Rev. 3

I0484572

The United States of America Fifth National Report for the Convention on Nuclear Safety

Manuscript Completed: August 2010
Date Published: September 2010

Office of Nuclear Reactor Regulation

ABSTRACT

The United States (U.S.) Nuclear Regulatory Commission (NRC) has updated NUREG-1650, Revision 2, "The United States of America Fourth National Report for the Convention on Nuclear Safety," issued September 2007, and will submit this report for peer review at the fifth review meeting of the Convention on Nuclear Safety at the International Atomic Energy Agency in Vienna Austria, in April 2011. This report addresses the safety of land-based commercial nuclear power plants in the U.S. It demonstrates how the U.S. Government achieves and maintains a high level of nuclear safety worldwide by enhancing national measures and international cooperation, and by meeting the obligations of all the articles established by the Convention. These articles address the safety of existing nuclear installations, the legislative and regulatory framework, the regulatory body, responsibility of the licensee, the priority given to safety, financial and human resources, human factors, quality assurance, assessment and verification of safety, radiation protection, emergency preparedness, siting, design and construction, and operation. Similar to the U.S. National Report issued in 2007, this revised document includes a section developed by the Institute of Nuclear Power Operations (INPO) describing work done by the U.S. nuclear industry to ensure safety. The prime responsibility for the safety of a nuclear installation rests with the license holder; therefore, Part 3 explains how the nuclear industry maintains and improves nuclear safety.

CONTENTS

EXECUTIVE SUMMARY

The U.S. Nuclear Regulatory Commission (NRC) has prepared Revision 3 to NUREG-1650, "United States of America Fifth National Report for the Convention on Nuclear Safety" for submission for peer review at the fifth review meeting of the Convention on Nuclear Safety, to be convened at the International Atomic Energy Agency in Vienna, Austria, in April 2011. The NRC issued the fourth report in September 2007. This revised report addresses the safety of land-based commercial nuclear power plants in the U.S. It demonstrates how the U.S. Government achieves and maintains a high level of nuclear safety worldwide by enhancing national measures and international cooperation and by meeting the obligations of all the articles established by the Convention. These articles address the safety of existing nuclear installations, the legislative and regulatory framework, the regulatory body, responsibility of the licensee, the priority given to safety, financial and human resources, human factors, quality assurance, assessment and verification of safety, radiation protection, emergency preparedness, siting, design and construction, and operation.

This report addresses the issues identified through the peer review conducted during the fourth review meeting in April 2008 and discusses challenges and issues that have arisen since that time. The fourth review meeting identified the following NRC challenges:

(1) hiring and developing a qualified workforce
(2) handling unexpected material degradation problems associated with operation and power up-rates
(3) maintaining a positive and adequate safety culture
(4) licensing new plants with new and different technologies

The NRC highlighted the following planned initiatives at the fourth review meeting:

(1) conduct follow-up activities related to the Integrated Regulatory Review Service (IRRS) self-assessment and prepare for the IRRS mission in 2010
(2) continue hiring and training initiatives
(3) continue the INPO 2-year evaluation program and the program to assist plants requiring additional support
(4) use the operating experience program to share experience and establish any underlying causes of unexpected material degradation
(5) complete the initiatives to establish the necessary framework to support the use of digital technology by drawing on the operating experience of others

This report also discussed the status of safety issues raised in the fourth U.S. National Report, including reactor materials degradation, unanticipated equipment problems from power uprates, and pressurized-water reactor (PWR) emergency core cooling system (ECCS) sump blockage resulting from post-loss-of-coolant accident (LOCA) chemical formation, as well as those that have arisen since 2007.

The Institute of Nuclear Power Operations (INPO) has also provided input to this report. The prime responsibility for the safety of a nuclear installation rests with the license holder; therefore, Part 3 explains how the nuclear industry maintains and improves nuclear safety.

PART 1

INTRODUCTION

This section describes the purpose and structure of the "United States of America Fifth National Report for the Convention on Nuclear Safety," the national policy of the U.S. toward nuclear activities, the main national nuclear programs, and current nuclear issues. It also highlights major regulatory accomplishments since submission of the previous (fourth) U.S. National Report in 2007 (see NUREG-1650, Revision 2, "The United States of America Fourth National Report for the Convention on Nuclear Safety," dated September 2007).

Purpose and Structure of This Report

The United States of America is submitting this updated report for peer review to the fifth review meeting of the Contracting Parties to the Convention on Nuclear Safety (hereafter referred to as the Convention or CNS). The scope of this report considers only the safety of land-based commercial nuclear power plants, consistent with the definition of nuclear installations provided in Article 2 and the scope of Article 3 of the Convention.

This report demonstrates how the U.S. Government meets the following objectives described in Article 1 of the Convention:

(i)   to achieve and maintain a high level of nuclear safety worldwide through the enhancement of national measures and international cooperation including, where appropriate, safety-related technical cooperation

(ii)   to establish and maintain effective defenses in nuclear installations against potential radiological hazards in order to protect individuals, society, and the environment from harmful effects of ionizing radiation from such installations

(iii) to prevent accidents with radiological consequences and to mitigate such consequences should they occur

Technical and regulatory experts from the U.S. Nuclear Regulatory Commission (hereafter referred to as the NRC, Commission,[1] agency, or staff) updated the fifth U.S. National Report, principally using agency information that is publicly available. This updated report follows the format of the fourth U.S. National Report, and is designed to be a stand-alone document. Hence, this report duplicates some of the information presented in the 2007 (fourth) report. To facilitate peer review, Part 1 of this report includes a summary of the main changes to the report (Table 1). This table is followed by a discussion of (1) the U.S policy towards nuclear activities, (2) national nuclear programs, (3) conclusions from the fourth review meeting, (4) current safety and regulatory issues, (5) an update on safety and regulatory issues discussed in the fourth U.S. National Report, (6) major regulatory accomplishments, and (7) the NRC's main challenges.

Part 2 discusses the Convention's Articles 6 through 19. Chapters are numbered according to the article of the Convention under consideration. Each chapter begins with the text of the article, followed by an overview of the material covered and a discussion of how the U.S. meets

---

[1]   "Commission" may also refer to the Chairman and Commissioners who head the NRC.

the obligations described in the article. Articles 6 through 9 summarize the existing nuclear installations and the legislative and regulatory system governing their safety and discuss the adequacy and effectiveness of that system. Articles 10 through 16 address general safety considerations and summarize major safety-related features. Articles 17 through 19 address the safety of installations.

Similar to the 2007 report, Part 3 of this document includes a contribution by INPO describing work done by the U.S. nuclear industry to ensure safety. INPO is a nongovernmental corporation founded in 1979 by the U.S. nuclear industry to collectively promote the highest levels of safety and reliability at U.S. nuclear plants. The prime responsibility for the safety of a nuclear installation rests with the license holder; therefore, Part 3 explains how the nuclear industry maintains and improves nuclear safety.

The report concludes with a series of appendices that discuss the NRC's main challenges as described in the NRC Strategic Plan and the Inspector General's report, followed by appendices of references, abbreviations, and acknowledgments. Annex 1 of the report lists nuclear plants in the U.S.

This report does not explicitly discuss Articles 1 through 5 because the general text of the report, and indeed the very existence of the report, fulfills the requirements of these articles. In accordance with Article 1, the report illustrates how the U.S. Government meets the objectives of the Convention. The report discusses the safety of nuclear installations according to the definition in Article 2 and the scope of Article 3. It addresses implementing measures (such as national laws, legislation, regulations, and administrative means) according to Article 4. Submission of this report fulfills the obligation under Article 5 on reporting. In addition, the information in this report is available in more detail on the NRC's public Web site.

Summary of Changes to the Fifth U.S. National Report

To facilitate peer review of this report, Table 1 summarizes the changes to the fifth U.S. National Report. A revision bar along the left margin of the page identifies changes from the fourth report.

Table 1   Summary of Changes to the Fifth U.S. National Report

Report Section	Change
Abstract	Updated to add discussion about the 4th CNS
Executive Summary	Updated to add discussion about the 4th CNS
PART 1	
Introduction	Updated to add discussion about the 4th CNS
Purpose and Structure of This Report	Updated to add discussion about the 4th CNS
Summary of Changes to the Fifth U.S. National Report	Updated table
The U.S. National Policy toward Nuclear Activities	Editorial changes only

4

Report Section	Change
National Nuclear Programs	Reordering and editorial changes only
Reactor Oversight Process	Updated to add discussion about the 2008 self-assessment
License Renewal	Updated to add discussion about units entering the 41st year of operation
Power Uprate Program	Editorial changes only
New Reactor Licensing	Updated to add discussion about applications received to date
Conclusions from the Fourth Review Meeting	Completely updated to add discussion about the 4th CNS
Items Resulting from Country Group Session	Completely updated to add discussion about the 4th CNS
Survey of Current Regulatory and Safety Issues	Completely updated to add discussion about seven current regulatory and safety issues
Reactor Materials Degradation Issues	New section
Cyber Security	New section
Digital Upgrades to Instrumentation and Control	New section
Moisture Effects on Underground Cables	New section
Containment Pressure Credit for Emergency Core Cooling System Pump Net Positive Suction Head	New section
Gas Voiding Issues in Light-Water Reactor Safety Systems	New section
Enhancements to Emergency Preparedness Regulations	New section
Status of Safety and Regulatory Issues Discussed in the Fourth U.S. National Report	Completely updated to add discussion about status of these issues discussed in the 4th National Report
Reactor Materials Degradation Issues	Completely updated section to add current status
Unanticipated Issues Associated with Power Uprates	Completely updated section to add current status
PWR Post-Loss of Coolant Accident Chemical Formation and PWR Sump Strainer Performance	Completely updated section to add current status
Other Major Regulatory Accomplishments	Added discussion about nine major regulatory accomplishments

Report Section	Change
Issuance of Early Site Permits and Limited Work Authorizations	Updated to add discussion about two early site permits and one limited work authorization issued
Reactor Pressure Vessel Pressurized Thermal Shock	New section
Power Reactor Security	New section
Aircraft Impact Assessment	New section
Fatigue Management	New section
Risk-Informed Fire Protection Infrastructure	New section
Probabilistic Risk Assessment Standard for the Analysis of External Events	New section
Regulatory Effectiveness	New section
Safety Culture Initiatives	New section
The NRC's Main Challenges	Updated to add discussion presented in the 2008-2013 NRC Strategic Plan
NRC Major Management Challenges	Updated to add discussion about the 2009 Inspector General's assessment
PART 2	
ARTICLE 6. EXISTING NUCLEAR INSTALLATIONS	Editorial changes only
6.1 Introduction	Updated to add safety strategic outcomes in fiscal years 2008-2009
6.2 Nuclear Installations in the United States	Updated to included 2009 reference
6.3 Regulatory Processes and Programs	Editorial changes only
6.3.1 Reactor Licensing	Updated to add discussion about applications received to date
6.3.2 Reactor Oversight Process	Updated to add discussion about the 2008 self-assessment
6.3.3 Industry Trends Program	Updated to add discussion about the baseline risk index for initiating events
6.3.4 Accident Sequence Precursor Program	Updated to include a discussion about the accident sequence precursor program status report issued in 2009
6.3.5 Operating Experience Program	Editorial changes only
6.3.6 Generic Issues Program	Updated to add discussion about changes to the program made in 2008
6.3.7 Rulemaking	Updated to add discussion about public access to rulemaking documents

Report Section	Change
6.3.8 Fire Regulation Program	Updated to add discussion about risk-informed, performance-based fire protection rule and the research program
6.3.9 Decommissioning	Updated to reference relevant regulations and guidance documents
6.3.10 Reactor Safety Research Program	Editorial changes only
6.3.11 Special Programs for Public Participation	Updated to add a discussion about the Federal Docket Management System
ARTICLE 7. LEGISLATIVE AND REGULATORY FRAMEWORK	Editorial changes only
7.1 Legislative and Regulatory Framework	Updated to add a list of ratified international conventions that impact nuclear safety
7.2 Provisions of the Legislative and Regulatory Framework	Editorial changes only
7.2.1 National Safety Requirements and Regulations	Updated to add discussion about regulations, executive orders and directives that impact nuclear safety
7.2.2 Licensing of Nuclear Installations	Updated to add discussion about the Atomic Energy Act, license renewal, and hearings
7.2.3 Inspection and Assessment	Updated to add discussion about resident inspectors
7.2.4 Enforcement	Updated monetary civil penalties limits and enforcement measures
ARTICLE 8. REGULATORY BODY	Editorial changes only
8.1 The Regulatory Body	Editorial changes only
8.1.1 Mandate	Editorial changes only
8.1.2 Authority and Responsibilities	Reorganized subsections
8.1.2.1 Scope of Authority	Editorial changes only
8.1.2.2 The NRC as an Independent Regulatory Agency	Updated to expand discussion
8.1.3 Structure of the Regulatory Body	Editorial changes only
8.1.3.1 The Commission	Editorial changes only
8.1.3.2 Component Offices of the Commission	Editorial changes only
8.1.3.3 Offices of the Executive Director for Operations	Noted organizational changes
8.1.3.4 Advisory Committees	Noted organizational changes
8.1.3.5 Atomic Safety and Licensing Board Panel	New section

Report Section	Change
8.1.4 International Responsibilities and Activities	Updated to expand discussion about treaties, export-import, assistance program, and international organizations
8.1.5 Financial and Human Resources	Editorial changes only
8.1.5.1 Financial Resources	Updated to add funds for fiscal years 2008-2010
8.1.5.2 Human Resources	Updated to expand discussion about recruitment, knowledge management and retaining staff
8.1.6 Position of the NRC in the Governmental Structure	Editorial changes only
8.1.6.1 Executive Branch	Editorial changes only
8.1.6.2 The States (i.e., of the United States)	Editorial changes only
8.1.6.3 Congress	Editorial changes only
8.1.7 Report of the Integrated Regulatory Review Service Self-Assessment Team	Updated to add discussion about complementary self-assessment performed in 2009 and 2010
8.2 Separation of Functions of the Regulatory Body from Those of Bodies Promoting Nuclear Energy	Editorial changes only
ARTICLE 9. RESPONSIBILITY OF THE LICENSE HOLDER	Editorial changes only
9.1 Introduction	Editorial changes only
9.2 The Licensee's Prime Responsibility for Safety	Editorial changes only
9.3 NRC Enforcement Program	Updated to reference revised guidance documents and discuss enforcement actions in 2008 and 2009
ARTICLE 10. PRIORITY TO SAFETY	Editorial changes only
10.1 Background	Updated to reference risk-informed, performance-based fire protection regulation
10.2 Probabilistic Risk Assessment Policy	Shortened
10.3 Applications of Probabilistic Risk Assessment	Updated discuss the use of RG 1.200
10.3.1 Risk-Informed Special Treatment	Revised title and updated to discuss the 50.69 final rule and RG 1.201, Revision 1
10.3.2 Risk Informed Inservice Inspection	Updated to add discussion about Code Case N-716

Report Section	Change
10.3.3 Risk-Informed Technical Specification Changes	Updated to expand discussion about accomplishments in the PRA area
10.3.4 Development of Standards	Updated to add discussion about a joint ASME/ANS PRA quality standard issued in 2009
10.4 Safety Culture	Editorial changes only
10.4.1 NRC Monitoring of Licensee Safety Culture	Editorial changes only
10.4.1.1 Background	Corrected section number
10.4.1.2 Enhanced Reactor Oversight Process	Corrected section number and updated to add discussion about the 2008 self-assessment
10.4.2 The NRC Safety Culture	Expanded discussion and added details about the Inspector General's 2009 survey.
10.5 Managing the Safety and Security Interface	New section
ARTICLE 11. FINANCIAL AND HUMAN RESOURCES	Editorial changes only
11.1 Financial Resources	Editorial changes only
11.1.1 Financial Qualifications Program for Construction and Operations	Editorial changes only
11.1.1.1 Construction Permit Reviews	Editorial changes only
11.1.1.2 Operating License Reviews	Editorial changes only
11.1.1.3 Combined License Application Reviews	Editorial changes only
11.1.1.4 Postoperating License Nontransfer Reviews	Editorial changes only
11.1.1.5 Reviews of License Transfers	Updated to add complete reference to NUREG-1577, Revision 1
11.1.2 Financial Qualifications Program for Decommissioning	Updated to add reference to 10 CFR 50.75
11.1.3 Financial Protection Program for Liability Claims Arising from Accidents	Updated Price-Anderson Act information
11.1.4 Insurance Program for Onsite Property Damages Arising from Accidents	Editorial changes only
11.2 Regulatory Requirements for Qualifying, Training, and Retraining Personnel	Editorial changes only
11.2.1 Governing Documents and Process	Editorial changes only
11.2.2 Experience	Shortened and updated numbers for human performance issues.

Report Section	Change
ARTICLE 12. HUMAN FACTORS	Editorial changes only
12.1 Goals and Mission of the Program	Editorial changes only
12.2 Program Elements	Updated to discuss the human event repository and analysis system
12.3 Significant Regulatory Activities	Editorial changes only
12.3.1 Human Factors Engineering Issues	Updated to reference NUREG-1852 and discuss the interim staff review guidance regarding computer-based procedures and plant digital upgrades
12.3.2 Emergency Operating Procedures and Plant Procedures	Updated experience subsection
12.3.3 Shift Staffing	Updated experience subsection
12.3.4 Fitness for Duty	Updated to add discussion about the fatigue management rulemaking and the Enforcement Guidance Memorandum
12.3.5 Human Factors Information System	Editorial changes only
12.3.6 Support to Event Investigations and For-Cause Inspections and Training	Updated to add discussion about safety culture inspections performed in 2007
ARTICLE 13. QUALITY ASSURANCE	Editorial changes only
13.1 Background	Editorial changes only
13.2 Regulatory Policy and Requirements	Section reworded
13.2.1 Appendix A to 10 CFR Part 50	Editorial changes only
13.2.2 Appendix B to 10 CFR Part 50	Editorial changes only
13.2.3 Approaches for Adopting More Widely Accepted International Quality Standards	Editorial changes only
13.3 Quality Assurance Regulatory Guidance	Updated slightly
13.3.1 Guidance for Staff Reviews for Licensing	Section renumbered and updated slightly.
13.3.2 Guidance for Design and Construction Activities	Section renumbered and updated slightly.
13.3.3 Guidance for Operational Activities	Section renumbered and updated slightly.
13.4 Quality Assurance Programs	Shortened and updated to discuss 10 CFR 52.103(g)
13.5 Quality Assurance Audits Performed by Licensees	New section
13.5.1 Audits of Vendors and Suppliers	New section
ARTICLE 14. ASSESSMENT AND VERIFICATION OF SAFETY	Editorial changes only

Report Section	Change
14.1 Ensuring Safety Assessments throughout Plant Life	Editorial changes only
14.1.1 Maintaining the Licensing Basis	Editorial changes only
14.1.1.1 Governing Documents and Process	Editorial changes only
14.1.1.2 Regulatory Framework for the Restart of Browns Ferry Unit 1	Editorial changes only
14.1.2 License Renewal	Editorial changes only
14.1.2.1 Governing Documents and Process	Updated to add discussion about revised guidance documents and rulemaking activities
14.1.2.2 Experience	Updated to add discussion about renewed license to date
14.1.2.3 Operating Beyond 60 Years	New section
14.1.3 The United States and Periodic Safety Reviews	Updated to expand discussion
14.1.3.1 The NRC's Robust and Ongoing Regulatory Process and the Current Licensing Basis	Editorial changes only
14.1.3.2 The Backfitting Process: Timely Imposition of New Requirements	Editorial changes only
14.1.3.3 The NRC's Extensive Experience with Broad-Based Evaluations	Updated to expand discussion about the Maintenance Rule
14.1.3.4 License Renewal Confirms Safety of Plants	Updated to expand discussion about the Reactor Oversight Process and license renewal
14.1.3.5 Risk-Informed Regulation and the Reactor Oversight Process	Updated to expand discussion about the Reactor Oversight Process
14.1.3.6 Licensee Responsibilities for Safety: Regulations and Initiatives Beyond Regulations	Editorial changes only
14.1.3.7 The NRC's Regulatory Process Compared with International Safety Reviews	Editorial changes only and removed figure.
14.2 Verification by Analysis, Surveillance, Testing, and Inspection	Updated to add discussion about performance measure and aging management
ARTICLE 15. RADIATION PROTECTION	Editorial changes only
15.1 Authorities and Principles	Editorial changes and updated to add discussion about new ICRP recommendations
15.2 Regulatory Framework	Editorial changes only

Report Section	Change
15.3 Regulations	Updated to add discussion about interaction with stakeholders and the evaluation of international standards
15.4 Radiation Protection Activities	Editorial changes only
15.4.1 Control of Radiation Exposure of Occupational Workers	Updated collective doses
15.4.2 Control of Radiation Exposure of Members of the Public	Updated to add background information about 10 CFR 20.1301 and 10 CFR 20.1302 and the revision of RGs 1.21 and 4.1
ARTICLE 16. EMERGENCY PREPAREDNESS	Editorial changes only
16.1 Background	Editorial changes only
16.2 Offsite Emergency Planning and Preparedness	Editorial changes only
16.3 Emergency Classification System and Emergency Action Levels	Editorial changes only
16.4 Recommendations for Protective Action in Severe Accidents	Updated number of States receiving potassium iodide and added reference to the draft revision to NUREG-0654/FEMA-REP-1, Supplement 3.
16.5 Inspection Practices - Reactor Oversight Process for Emergency Preparedness	Editorial changes only
16.6 Responding to an Emergency	Updated to add discussion about the National Response Framework issued in 2008
16.6.1 Federal Response	Updated to add discussion about updates to governing documents
16.6.2 Licensee, State, and Local Response	Editorial changes only
16.6.3 The NRC's Response	Updated to expand discussion about response centers
16.6.4 Aspects of Security that Support Response	Editorial changes and updated to add reference to rulemaking discussions
16.7 International Arrangements	Updated to add renewal dates of bilateral agreements
ARTICLE 17. SITING	Editorial changes only
17.1 Background	Updated to add discussion about applications received to date
17.2 Safety Elements of Siting	Editorial changes only
17.2.1 Background	Editorial changes only

Report Section	Change
17.2.2 Assessments of Seismic and Geological Aspects of Siting	Updated to add discussion about seismic designs in new reactors
17.2.3 Assessments of Radiological Consequences	Editorial changes only
17.3 Environmental Protection Elements of Siting	Editorial changes only
17.3.1 Governing Documents and Process	Updated to add discussion about changes in review practices made in 2007 and 2010 and added discussion about the memorandum of understanding between the NRC and the U.S. Army Corps of Engineers.
17.3.2 Other Considerations for Siting Reviews	Editorial changes only
17.4 Consultation with other Contracting Parties to be Affected by the Installation	New section
ARTICLE 18. DESIGN AND CONSTRUCTION	Editorial changes only
18.1 Defense-in-Depth Philosophy	Editorial changes only
18.1.1 Governing Documents and Process	Editorial changes only
18.1.2 Experience	Editorial changes only
18.1.2.1 Regulatory Framework for the Reactivation of Watts Bar Unit 2	Updated status of the reactivation
18.1.2.2 Design Certifications	Updated to add discussion about applications received to date
18.2 Technologies Proven by Experience or Qualified by Testing or Analysis	Editorial changes only
18.3 Design for Reliable, Stable, and Easily Manageable Operation	Editorial changes only
18.3.1 Governing Documents and Process	Updated references
18.3.2 Experience	Editorial changes only
18.3.2.1 Human Factors Engineering	New section
18.3.2.2 Digital Instrumentation and Controls	Renumbered section and updated.
18.3.2.3 Cyber Security	Renumbered section and updated to add discussion about new regulations and guidance documents.
18.4 New Reactor Construction Experience Program	New section
ARTICLE 19. OPERATION	Editorial changes only

Report Section	Change
19.1　Initial Authorization to Operate	Shorten and reorganized; updated to include discussion of applications received to date.
19.2　Definition and Revision of Operational Limits and Conditions	Editorial changes only
19.3　Approved Procedures	Added references
19.4　Procedures for Responding to Anticipated Operational Occurrences and Accidents	Editorial changes only
19.5　Availability of Engineering and Technical Support	Editorial changes only
19.6　Incident Reporting	Updated to add discussion about abnormal occurrence report to Congress, the International Nuclear and Radiological Event Scale and the nuclear events Web-based system
19.7　Programs To Collect and Analyze Operating Experience	Updated to expand discussion about the phases of the Operating Experience Program and international operating experience
19.8　Radioactive Waste	Updated to add the status of the high-level waste repository in Nevada
PART 3	
Convention on Nuclear Safety Report: The Role of the Institute of Nuclear Power Operations in Supporting the U.S. Commercial Nuclear Power Industry's Focus on Nuclear Safety	Updated
APPENDIX A　NRC STRATEGIC PLAN 2008-2013	Updated to add new Strategic Plan
APPENDIX B　NRC MAJOR MANAGEMENT CHALLENGES FOR THE FUTURE	Updated to add the 2009 report from the Inspector General
APPENDIX C　REFERENCES	Updated
APPENDIX D　ABBREVIATIONS	Updated
APPENDIX E　ACKNOWLEDGMENTS	Updated
ANNEX 1　U.S. COMMERCIAL NUCLEAR POWER REACTORS	Updated
ANNEX 2　U.S. NUCLEAR ELECTRIC INDUSTRY PERFORMANCE INDICATOR GRAPHS	New section.　Graphs moved from Part 3 to Annex 2 to maintain consistency in the report.

The U.S. National Policy toward Nuclear Activities

The Energy Reorganization Act of 1974 created the NRC as an independent agency of the Federal Government.   The agency's mission is to license and regulate the Nation's civilian use of byproduct, source, and special nuclear materials to ensure adequate protection of public health and safety, promote the common defense and security, and protect the environment. The agency also has a role in combating the proliferation of nuclear materials worldwide.   The NRC's safety and security responsibilities stem from the Atomic Energy Act of 1954, as amended.   The agency accomplishes its mission by licensing and overseeing nuclear reactor operations and other activities that apply to the possession of nuclear materials and wastes, ensuring that nuclear materials and facilities are safeguarded from theft and radiological sabotage, issuing rules and standards, inspecting nuclear facilities, and enforcing regulations.

The NRC, in conducting its work, adheres to seven organizational values to guide its actions: integrity, service, openness, commitment, cooperation, excellence, and respect.   The principles of good regulations help carry out the NRC regulatory activities.   These principles focus on ensuring safety and security while appropriately balancing the interests of stakeholders, including the public and licensees.   These principles are independence, efficiency, clarity, reliability and openness.   The NRC's final decisions are based on objective, unbiased assessments of all information, and are documented with reasons explicitly stated.   The NRC establishes means to evaluate and continually upgrade its regulatory capabilities.   Its regulations are coherent, logical, practical, and based on the best available knowledge from research and operational experience.

The NRC also views nuclear regulation as the public's business and, as such, it must be transacted openly and candidly to maintain the public's confidence.   The NRC issuance of its Open Government Plan, dated June 7, 2010, is a reflection of its long history of, and commitment to, openness with the public and transparency in the regulatory process.   The agency's goal to ensure openness explicitly recognizes that the public must be informed about, and have a reasonable opportunity to participate meaningfully, in the regulatory process.   Except for certain proprietary business material, facility safeguards information, sensitive pre-decisional information, and information supplied by foreign countries that is deemed to be sensitive, the NRC makes the documentation that it uses in its decision-making process available in the agency's Public Document Room in Rockville, MD, and on the agency's public Web site at http://www.nrc.gov.   As a result, a significant amount of information about nuclear activities and the national policy regarding them is available to everyone.

The NRC's interpretation of regulations continues to evolve from a prescriptive, deterministic approach toward a more risk-informed and performance-based regulatory approach.   Improved probabilistic risk assessment (PRA) techniques, combined with more than four decades of accumulated experience with operating nuclear power reactors, led the Commission to revise or eliminate certain requirements.   The Commission is also prepared to strengthen the regulatory system when risk considerations reveal the need.

National Nuclear Programs

The NRC has a number of programs and processes to protect public health and safety and the environment and to meet the obligations of the Convention.   Key programs and processes in the reactor arena comprise a well-established licensing process, which includes: (1) reactor oversight, (2) license renewal, (3) power uprates, and (4) new reactor licensing.

Reactor Oversight Process

The NRC's Reactor Oversight Process is now nearly 9 years old. In its annual self-assessment for calendar year 2008, the NRC staff concluded that the Reactor Oversight Process provided effective safety oversight as demonstrated by meeting the program goals and achieving its intended outcomes. The self-assessment showed that the Reactor Oversight Process was objective, risk-informed, understandable, and predictable. It also showed that the Reactor Oversight Process ensures openness and effectiveness in support of the agency's mission and its strategic goals of safety and security. The NRC appropriately monitored operating nuclear power plant activities and focused agency resources on performance issues. Plants continued to receive a level of oversight commensurate with their performance. The staff continued to emphasize stakeholder involvement and improve various aspects of the Reactor Oversight Process as a result of feedback and lessons learned.

Article 6 of this report discusses the Reactor Oversight Process in detail.

License Renewal

The NRC's review of license renewal applications focuses on maintaining plant safety and particularly considers the effects of aging on important structures, systems, and components. The review of a renewal application proceeds along two paths—one to review safety issues and the other to assess potential environmental impacts. Applicants must demonstrate that they have identified and can manage the effects of aging and can continue to maintain an acceptable level of safety throughout the period of extended operation. Applicants must also address the environmental impacts from extended operation. With the improved economic conditions for operating nuclear power plants, the Commission has seen sustained, strong interest in license renewal, which allows plants to operate up to 20 years beyond their current operating licenses. The Atomic Energy Act established the original 40-year term, which was not based on technical limitations.

The decision to seek license renewal is voluntary and rests entirely with nuclear power plant owners. The decision is typically based on the plant's economic viability and whether it can continue to meet the Commission's requirements. Currently, more than half of the plants in the United States have had their operating licenses renewed. Based on statements from industry representatives, the Commission expects nearly all sites to apply for license renewal. In 2009, four units entered their 41st year of operation. These were Oyster Creek (April), Nine Mile Point Unit 1 (August), Ginna (September), and Dresden Unit 2 (December). In 2010, three additional units enter the period of extended operation. These units are H. B. Robinson Unit 2 (July), Monticello Unit 1 (September), and Point Beach Unit 1 (October.)

Article 14 of this report discusses the license renewal process in detail.

Power Uprate Program

Under its licensing program, the NRC carefully reviews requests to raise the maximum thermal power level at which a plant may be operated. In reviewing these power uprate requests,

NRC's review focuses on safety. The agency closely monitors operating experience to identify safety issues that may affect the implementation of power uprates.

Power uprates can be classified as: (1) measurement uncertainty recapture power uprates, (2) stretch power uprates, and (3) extended power uprates (EPUs). Measurement uncertainty recapture power uprates are less than a two-percent increase and are achieved by implementing enhanced techniques for calculating reactor power. Stretch power uprates are typically increases of up to seven percent and are generally within the original design capacity of the plant. Stretch power uprates usually involve changes to instrumentation setpoints and do not generally involve major plant modifications. EPUs are usually greater than stretch power uprates and require significant modifications to major balance-of-plant equipment. The NRC has approved EPUs of up to 20 percent.

New Reactor Licensing

The NRC staff is engaged in numerous ongoing interactions with vendors and utilities regarding prospective new reactor applications and licensing activities. Based on these interactions, the NRC staff has received a significant number of new reactor combined license applications since 2007. As of March 1, 2010, the NRC has received 18 combined license applications for 28 new light-water reactor units. Of these 18 applications, five applicants have requested that the NRC suspend its review of their applications given changing business strategies. The NRC is now actively reviewing 13 combined license applications. All combined license applicants are using the licensing process specified in the recently revised Title 10 of the *Code of Federal Regulations* (10 CFR) Part 52, "Licenses, Certifications, and Approvals for Nuclear Power Plants," which is designed to be more stable and predictable than the process specified in 10 CFR Part 50, "Domestic Licensing of Production and Utilization Facilities." This licensing process resolves all safety and environmental issues, as well as emergency preparedness and security issues, before a new nuclear power plant is constructed.

The NRC staff has issued design certifications for four reactor designs that can be referenced in an application for a nuclear power plant: (1) General Electric (GE) Nuclear Energy's Advanced Boiling Water Reactor (ABWR), (2) Westinghouse Electric Company, LLC's (Westinghouse's) System 80+, (3) Westinghouse's Advanced Passive (AP) 600 design, and (4) Westinghouse's AP1000.

The NRC staff is currently performing the following design certification reviews: (1) GE-Hitachi Nuclear Energy's Economic Simplified Boiling Water Reactor (ESBWR), (2) Westinghouse's AP1000 design amendment, (3) AREVA Nuclear Power's U.S. Evolutionary Power Reactor (US EPR), (4) Mitsubishi Heavy Industries, Ltd.'s U.S. Advanced Pressurized Water Reactor (US APWR), and (5) South Texas Project Nuclear Operating Company's ABWR application to address the aircraft impact rule.

By certifying nuclear reactor designs, the NRC resolves safety issues in a design certification rulemaking. When an applicant submits an application for construction of a new nuclear power plant using one of the certified designs, the license application review can proceed more efficiently in a manner that ensures safety while minimizing unnecessary regulatory burden and delays.

To date, the NRC has issued four early site permits: (1) System Energy Resources, Inc., for the

Grand Gulf site in Mississippi; (2) Exelon Generation Company, LLC, for the Clinton site in Illinois; (3) Dominion Nuclear North Anna, LLC, for the North Anna site in Virginia; and (4) Southern Nuclear Operating Company for the Vogtle Electric Generating Plant early site permit and limited work authorization in Georgia. These are the first early site permits issued by the NRC and the first time this portion of the 10 CFR Part 52 licensing process has been implemented. According to this process, environmental issues that have been resolved in the early site permit proceedings cannot be re-opened during a combined license proceeding.

By letter dated July 1, 2009, Exelon notified the NRC staff that Exelon had decided to pursue an early site permit rather than a combined license for the Victoria station in Texas. By letter dated October 13, 2009, Exelon notified the NRC staff its plan to submit an early site permit application in late March 2010. On March 25, 2010, Exelon submitted its early site permit application for Victoria station. The application uses the plant parameter envelope approach for two units, includes a complete emergency plan, and did not request a limited work authorization. In a letter dated June 8, 2010, the staff informed Exelon that the Victoria station early site permit application was accepted for docketing. At this time, the staff is developing a technical review schedule. The safety and environmental reviews are planned to begin in October 2010.

By letter dated December 2, 2008, Public Service Enterprise Group updated the NRC staff on its intention to submit an application for an early site permit during the second quarter of calendar year 2010. Public Service Enterprise Group's early site permit application was submitted on May 25, 2010. The application uses a plant parameter envelope methodology because a reactor technology has not been selected yet. On August 5, 2010, the staff completed its acceptance review and docketed the application.

In 2006, to better prepare the agency for the anticipated new reactor licensing and construction inspection work, while ensuring that the agency maintains its focus on the safety and security of currently operating reactors, the NRC established the Office of New Reactors. The agency also established a dedicated construction inspection organization in its Region II office in Atlanta, Georgia, that will carry out all construction inspection activities across the U.S., including both the day-to-day onsite inspections and the specialized inspections needed to support NRC oversight of the construction of new nuclear power plants.

One partially built plant, Watts Bar Nuclear Plant Unit 2, had stopped construction activities in the mid-1980s. Watts Bar Unit 2 is a Westinghouse designed PWR located in southeastern Tennessee and owned by the Tennessee Valley Authority (TVA), which has resumed construction activities and is currently pursuing an operating license approval under 10 CFR Part 50.

In addition to working on domestic issues for new reactor construction, the NRC has been a leader in cooperating with other national nuclear regulatory authorities to address advanced reactor oversight. The NRC is participating in an international effort, the Multinational Design Evaluation Program, to more efficiently review new reactor designs. The goal of this program is to make all new reactor reviews more safety-focused. NRC representatives are communicating closely with representatives from the Finnish and French regulatory authorities concerning the European power reactor designs that are under construction in Finland and slated to be licensed in France and the United States. The NRC is also participating in a longer-term multinational effort to establish reference regulatory practices and regulations for the review of current and future reactor designs.

Articles 17 and 18 of this report discuss the new reactor licensing in more detail.

Conclusions from the Fourth Review Meeting

This section presents the conclusions from the review of the 2007 U.S. National Report at the fourth review meeting in April 2008.

Delegates from other countries noted that the U.S. delivered a highly informative presentation at the country group meeting.   They commended the U.S. for including a contribution from INPO in the report that explains how the nuclear industry maintains and improves nuclear safety.

Items Resulting from Country Group Session

Review of the questions raised by other contracting parties on the U.S. National Report identified the following areas of interest:

- safety trends
- generic issues
- long-term operation
- new and advance reactors
- knowledge management
- regulatory openness

The NRC's presentation during the 2008 review meeting focused on these topics.   INPO also discussed its role in maintaining and improving nuclear safety.

Country Group 1 participants concluded that the U.S. implemented the following good practices:

- the National Report content and structure
- involving the industry in the development of the National Report and the review meeting presentation
- making extensive use of the NRC public Web site to increase public awareness
- establishing the Office of New Reactors and hiring staff in advance of new reactor construction
- developing the new reactors licensing structure
- performing a self-assessment in preparation for the 2010 Integrated Regulatory Review Service (IRRS) mission

Country Group 1 identified the following challenges for the U.S.:

- hiring and developing a qualified workforce
- handling unexpected material degradation problems associated with plant operation and power up-rates
- maintaining a positive and adequate safety culture
- licensing new plants with new and different technologies

Country Group 1 highlighted the following planned U.S. initiatives:

- conduct follow-up activities related to the IRRS self assessment and prepare for the IRRS mission in 2010
- continue hiring and training initiatives
- continue the INPO 2-year evaluation program and the program to assist plants requiring additional support
- use the operating experience program to share experience and establish any underlying causes of unexpected material degradation
- complete the initiatives to establish the necessary framework to support the use of digital technology drawing on the operating experience of others

The current U.S. National Report addresses many of these issues under the relevant articles.

Survey of Current Safety and Regulatory Issues

The NRC and its licensees are currently facing the following safety and regulatory issues:

- reactor materials degradation
- cyber security
- digital upgrades to instrumentation and control
- moisture effects on underground cables
- containment pressure credit for emergency core cooling system pump net positive suction head
- gas voiding impacts on emergency core cooling system operability
- proposed changes to emergency preparedness regulations

Reactor Materials Degradation Issues

Cases involving materials degradation include the degradation of buried piping systems and the degradation of neutron-absorber materials in spent fuel pools.

Degradation of Buried Piping Systems

Over the past several years, instances of buried piping leaks have occurred in safety-related and nonsafety-related piping at nuclear power plants. Most of the leaks have occurred in nonsafety-related piping. Some of these leaks have caused inadvertent releases of low-level radioactive material and diesel fuel oil. This has resulted in ground water contamination at several plants. The pipe degradation leading to these leaks has not affected the operability of safety systems, and the type and amount of radioactive material or chemicals released to the environment have been a small fraction of the regulatory limits. Consequently, these pipe leaks have been of low significance with respect to public health and safety. The staff documented its evaluation of buried piping degradation issues in SECY-09-0174, "Staff Progress in Evaluation of Buried Piping at Nuclear Reactor Facilities," dated December 2, 2009.

Based on the staff's review, including the review of operating experience related to buried piping degradation, current regulations and American Society of Mechanical Engineers (ASME) Boiler and Pressure Vessel Code requirements are effective in ensuring that the structural integrity and functionality of buried, safety-related piping are maintained. Current regulations are also effective in ensuring that unintended releases of hazardous material to the environment from leaks in both safety-related and nonsafety-related buried piping remain below regulatory limits.

The U.S. nuclear industry has recently developed the Buried Piping Integrity Initiative. The staff plans to meet with the industry to further understand this initiative, evaluate its effectiveness, and monitor industry implementation. The staff will evaluate the need to revise NRC inspection procedures to assess licensee implementation of this new initiative. The staff will also continue to actively participate in codes and standards activities, revise license renewal guidance, monitor operating experience, and assess the need for any further regulatory actions or communications.

In addition, in March 2010 the NRC established a task force to evaluate its regulatory framework associated with groundwater protection. The objective of the task force was to evaluate NRC actions to date addressing buried piping leaks and whether those actions needed to be augmented. The report "Groundwater Task Force Final Report," dated June 2010, documents the task force's observations, conclusions, and recommendations in a number of areas, including policy and communications. Currently, a Senior Management Review Group has been formed from a group of NRC senior executives, and has been tasked to decide whether the NRC agrees with the findings of the task force and how best to act upon the conclusions and recommendations contained in the final report.

Degradation of Neutron-Absorber Materials in Spent Fuel Pools

One of the NRC's strategic outcomes for its safety goal is that there are "no inadvertent criticality events." To achieve this goal, as it relates to the storage and handling of reactor fuel, the NRC has promulgated regulations focused on maintaining spent fuel pools subcritical under normal and accident conditions. These regulations appear in 10 CFR 50.68, "Criticality Accident Requirements," and General Design Criterion 62, "Prevention of Criticality in Fuel Storage and Handling," in 10 CFR Part 50, Appendix A. To satisfy these regulations, most licensees have installed fixed neutron absorbers within the spent fuel pool storage racks. Degradation or deformation of the credited neutron absorbing materials could reduce the material's ability to perform its safety function and potentially violate the NRC's subcriticality regulations.

There are many different types of neutron absorbing materials. Within U.S. spent fuel pools the most common types are Boraflex, carborundum, boral, and Metamic. Boraflex was the first neutron-absorbing material to exhibit significant degradation. The NRC documented this issue in Information Notice (IN) 87-43, "Gaps in Neutron-Absorbing Material in High- Density Spent Fuel Storage Racks," dated September 8, 1987; IN 93-70, "Degradation of Boraflex Neutron Absorber Coupons," dated September 10, 1993; and IN 95-38, "Degradation of Boraflex Neutron Absorber in Spent Fuel Pool Storage Racks," dated September 8, 1995; and in Generic Letter (GL) 96-04, "Boraflex Degradation in Spent Fuel Pool Storage Racks," dated June 26, 1996. Ultimately, this issue was resolved through either revised plant-specific criticality analyses that reduced or eliminated credit for Boraflex or by the replacement of Boraflex with other neutron-absorbing materials.

Recent operating experience has identified several instances of degradation, deformation, or both of carborundum and boral neutron-absorbing materials in the spent fuel pools of operating reactors. One example of neutron-absorbing material degradation occurred in the Palisades Power Plant. On July 15, 2008, in support of its license renewal activities, the licensee performed "blackness testing" of the spent fuel pool racks to verify its carborundum was performing in accordance with the assumptions in its criticality analysis of record. Based on this testing, the licensee could not confirm that the spent fuel pool met the subcriticality requirements in 10 CFR 50.68 or its technical specifications. Since the licensee did not have an established monitoring program for the carborundum, the onset of the degradation and the degradation rate cannot be established. In response to the recent operating experience on this issue, the NRC issued IN 09-26, "Degradation of Neutron-Absorbing Materials in the Spent Fuel Pool," dated October 23, 2009, and Draft License Renewal Interim Staff Guidance (ISG) 2009-01, "Staff Guidance Regarding Plant-Specific Aging Management Review and Aging Management Program for Neutron-Absorbing Material in Spent Fuel Pools," dated November 23, 2009. The NRC is currently working to finalize the ISG and exploring what additional actions need to be taken.

The NRC has begun to evaluate the regulatory changes that may be necessary to ensure that its licensees can identify and mitigate neutron-absorber degradation before it challenges subcriticality safety margins. The Palisades operating experience has highlighted the importance of an effective surveillance program for the early identification of neutron-absorber degradation. Such a program could consist of various testing and identification methods, including, but not necessarily limited to, coupon sampling, in-situ testing, and validated and verified predictive analytical computer codes.

Cyber Security

Information security programs continue to be a critical consideration for any organization that depends on information systems and computer networks to carry out mission or business objectives. The energy sector and the necessary regulatory activities within that sector to provide safe power generation are not immune to increasing threats to their information management and computer enabled control systems. These threat vectors include: cyber criminals, unauthorized access, insider misuse, denial of service attacks, natural disasters, and other disruptions.

Over the last few years, power generators have markedly increased their use of digital control systems to regulate, monitor, and operate power production facilities. This increase in the use of digital control systems has been more than matched by the recent increase in security incidents reported both domestically and internationally, the ease with which computer hacking tools are available, and the steady advancement in the sophistication and effectiveness of attack technology. These risks and risk vectors have all contributed to the urgency of power generators and regulators to ensure that this infrastructure is supported and protected by strong, effective, and measurable information systems security programs.

In March 2009, the NRC issued a new rule on cyber security, 10 CFR 73.54, "Protection of Digital Computer and Communication Systems and Networks." This rule requires operating power reactor licensees and combined operating license applicants to provide assurance that nuclear power plants' safety, safety-related, security, and emergency preparedness functions are protected from cyber attacks up to and including the design-basis threat. This new regulation required licensees and combined operating license applicants to submit a cyber security plan, including an implementation schedule, to the NRC for review and approval.

In addition, the NRC has taken steps to address these issues within the agency by establishing the Computer Security Office.

Section 18.3.2.3 of this report discusses cyber security in more detail.

Digital Upgrades to Instrumentation and Control

The use of digital instrumentation and control raises issues that were not relevant to analog systems. Examples of such issues include the following:

- A common-cause failure attributable to software errors was not possible with analog systems. This potential weakness may require the consideration of diversity and defense-in-depth in the application of digital instrumentation and control systems.

- Interchannel communication, communication between nonsafety and safety systems, and system security and reliability must be reviewed closely to ensure that public safety is preserved.

- Highly integrated control room designs with safety and nonsafety displays and controls will be the norm for new reactor designs.

- Human factors design and quality assurance during all phases of software development, control, and validation and verification are critical.

The NRC's Digital Instrumentation and Control Steering Committee initiated task working groups to develop ISG documents for all high-priority technical issues associated with licensing digital instrumentation and control for nuclear power reactors. The working groups developed the ISG documents with significant input from external stakeholders through a series of public meetings and posted draft versions on the NRC Web site for public comment. The working groups addressed the following technical issues: (1) cyber security, (2) diversity and defense-in-depth, (3) review of new reactor digital instrumentation and control PRA, (4) highly-integrated control room-communications, and (5) highly-integrated control room human factors. The NRC staff is using the guidance documents to conduct ongoing reviews. Early feedback from licensees and NRC staff who have used the ISG documents has been positive. The staff used the ISGs in reviewing digital upgrades for the Wolf Creek and Oconee plants and in reviewing a number of design certification and combined license for new plants. The NRC staff plans to use the ISGs to update regulatory documents such as standard review plans (SRPs), regulatory guides (RGs), and NUREGs.

The working groups are still developing guidance on the licensing process for operating power reactors and fuel cycle facilities. For the licensing process, the working group is providing additional guidance on the scope and conduct of the review of digital retrofits to operating plant safety systems. The staff is incorporating lessons learned from ongoing reviews and has published a draft ISG for comment. For fuel cycle facilities, the working group is addressing many of the same technical and licensing questions, with consideration of the consequences of digital system failures and how they are treated by the significant differences in risk profiles and licensing requirements for power plants and fuel cycle facilities.

The NRC is actively involved with the Multinational Design Evaluation Program which is an international assembly of nuclear regulators addressing common issues with the licensing of new reactors. The NRC chairs the digital instrumentation and control issue-specific group, which is looking at ways to harmonize requirements, standards, and guidance for instrumentation and control. The NRC is also working with the EPR digital instrumentation and control task group, which is a collaboration of regulators that are reviewing the EPR instrumentation and control design. The Multinational Design Evaluation Program allows the NRC to share digital instrumentation and control information to support regulatory infrastructure improvements and licensing decisions.

Article 18 of this report discusses the digital instrumentation and control in more detail.

Moisture Effects on Underground Cables

The NRC began a detailed review of underground electrical power cables after an increasing trend in moisture-induced cable failures was identified. The failed cables had been exposed to condensation, wetting, submergence, and other environmental stresses that resulted in insulation degradation. Since most of the cables exposed to this environment were not designed for continuous wetting or submergence, there is an increasing possibility of multiple failures, which in turn could initiate a plant shutdown and/or disable accident mitigation systems.

On February 7, 2007, the NRC issued GL 2007-01, "Inaccessible or Underground Power Cable Failures That Disable Accident Mitigation Systems or Cause Plant Transients," to inform licensees that the failure of certain power cables can affect the functionality of multiple accident mitigation systems or cause plant transients. The NRC asked the licensees to provide information on inaccessible or underground power cable failures for all cables that are within the scope of 10 CFR 50.65, "Requirements for Monitoring the Effectiveness of Maintenance at Nuclear Power Plants" (the Maintenance Rule).

Based on the review of licensee's responses to GL 2007-01, the NRC staff identified 269 cable failures at U.S. nuclear power plants. Licensees applying for a 20-year license renewal have agreed to implement a cable testing program during the period of extended operation for a limited number of cables that are within the scope of licensee renewal, but only a few plants have established a cable testing program for the current operating period. The data obtained from the responses to GL 2007-01 show an increasing trend in cable failures within the plants' current 40-year licensing period of operation. The predominant factor contributing to cable failures at nuclear power plants appears to be the presence of water or moisture resulting in intrusion, because of the submergence of underground cables in water. If cables have been exposed to conditions for which they are not designed, licensees need to demonstrate, through adequate testing, that there is reasonable assurance that the cables can perform their intended design function. Licensees should also minimize the amount of moisture in underground cable raceways, conduits, and cable vaults.

NRC regulations in 10 CFR Part 50 require licensees to assess the condition of systems and components in a manner sufficient to provide reasonable assurance that they are capable of fulfilling their intended functions, and that a test program to ensure that components will perform satisfactorily in service is identified and performed. Licensees should have a program for using available diagnostic cable testing methods to assess cable condition to ensure the insulation is not degraded over the life of the plant.

In January 2010, the NRC issued NUREG/CR-7000, "Essential Elements of an Electric Cable Condition Monitoring Program," to inform licensees of the types of cable testing methods that are currently available to detect cable insulation degradation. In addition, the Electric Power Research Institute (EPRI) has also developed a model cable monitoring program to provide licensees with information on creating such a program. In June 2010, the NRC staff issued draft guidance DG-1240, "Condition Monitoring Program for Electric Cables Used in Nuclear Power Plants," for public comment. The comment period ended on August 13, 2010. Currently, the staff is evaluating the comments received. The staff expects to issue the final RG by January 2011.

Containment Pressure Credit for Emergency Core Cooling System Pump Net Positive Suction Head

NRC RG 1.1, "Net Positive Suction Head for Emergency Core Cooling and Containment Heat Removal System Pumps," dated November 2, 1970, states that the pressure in containment before the postulated accident should be used when determining the available net positive suction head of emergency core cooling system and containment heat removal system pumps. Before the NRC issued this guidance document, some reactors were designed and licensed using the calculated containment accident pressure.

The agency modified this guidance in RG 1.82, Revision 3, "Water Sources for Long-Term Recirculation Cooling Following a Loss of Coolant Accident," dated November 2003, which permitted certain operating reactors to use containment accident pressure when modification of the reactor design was impractical. The modification to the guidance of RG 1.1 recognized the fact that in certain cases it was not practical to avoid using containment accident pressure. Such cases included sub-atmospheric containments, application of a larger debris source term following a loss-of-coolant accident, and an increase in licensed thermal power (or power uprates).

As a result of discussions with the NRC Advisory Committee on Reactor Safeguards, the staff is re-examining this issue with the goal of evaluating containment integrity probabilistically and studying all related pump phenomena and quantifying margins both in terms of pump cavitation limits and containment accident response. Some of the subjects examined include the effect of containment integrity testing frequency on failure probabilities, the uncertainty in required net positive suction head, cavitation erosion as a function of pump flow rate, and the mechanical performance of centrifugal pumps with various degrees of cavitation. The staff is also evaluating whether this issue raises a policy question regarding the use of probabilistic risk assessment in deterministic regulatory decisionmaking and defense-in-depth. The staff will publish the results of this work in appropriate regulatory documents.

Gas Voiding Issues in Light-Water Reactor Safety Systems

The accumulation of gas in systems that are important to safety has been a continuing, often unrecognized, problem since the first light-water nuclear power plants were placed into operation. Early manifestations of the issue included pipe hanger damage as a result of water hammer in residual heat removal systems when the systems were started and the loss of residual heat removal when the pumps became gas-bound. This led to a recognition of potential problems with the emergency core cooling systems since much of the residual heat removal system also serves as the low-pressure – high flow rate portion of the emergency core cooling system, and similar problems could occur in the low-pressure and high-pressure emergency core cooling systems if they were placed in operation in response to a loss-of-coolant accident. Consequently, numerous publications were issued to address the issue, technical specifications were developed to require pump discharge piping to be full of water to address the water hammer issue, and steps were taken to prevent gas ingestion into pumps. Before 2008, the actions were not fully successful because of a failure to understand the root causes of gas accumulation and to comprehensively address the potentially affected systems and the phenomena associated with gas accumulation and movement before, during, and after system startup.

The root causes of gas accumulation include: (1) designs that allow gas introduction and accumulation, (2) licensees failing to properly fill and vent the system following drain-down or maintenance, (3) ineffective gas accumulation controls during operation, (4) inappropriate technical specifications regarding the scope and frequency of inspections for gas accumulation, and (5) unanticipated problems with keep-full systems.

GL 2008-01, "Managing Gas Accumulation in Emergency Core Cooling, Decay Heat Removal and Containment Spray Systems," dated January 11, 2008, addressed the issue for several important safety systems via in-depth coverage of the phenomena and the operating processes necessary to prevent event occurrence as a result of gas. The U.S. nuclear industry provided a detailed response to GL 2008-01 that included: (1) suction pipe testing, (2) development of analysis methodologies, (3) system walkdowns, including precision measurement of piping configurations, (4) void measurements using ultra-sonics, rewritten and new procedures, (5) extensive operator training, and (6) hardware changes such as the addition of vent valves and tanks to remove gas from piping before it becomes a concern. These follow-up actions have resulted in an enhanced understanding of the issues and implementation of measures to minimize future problems. As a result, there is an increased confidence that the systems will perform their safety-related functions when required to do so. Further improvements are underway. These include the development of improved void behavior analysis methods, increased in-depth coverage of transient behavior during pump starts, improved technical specification coverage including surveillance requirements, increased technical coverage including systems that were not identified in GL 2008-01, and improvements in plant operation including areas such as the corrective action plan, procedures, and operator training.

The NRC is following up on the industry activities by reviewing licensee responses to GL 2008-01 and by performing inspections at the 104 nuclear power plants that are licensed in the U.S. The scope of these activities is illustrated by the generic review instructions the NRC uses in providing inspection suggestions to its inspectors in accordance with Temporary Instruction 2515/177, "Managing Gas Accumulation in Emergency Core Cooling, Decay Heat Removal, and Containment Spray Systems (NRC Generic Letter 2008-01)," dated June 9, 2009. The scope of industry participation is evident in the four well-attended workshops sponsored by the Nuclear Energy Institute (NEI) and in the release of NEI guidance document NEI 09-10, Revision 0, "Guidelines for Effective Prevention and Management of System Gas Accumulation," dated October 2009.

Enhancements to Emergency Preparedness Regulations

The basis of radiological emergency preparedness and response is to protect public health and safety by avoiding public radiological exposure as a result of a release from a nuclear power plant. Since the Three Mile Island accident in 1979, the premise underlying emergency preparedness regulations has been that conditions and events driving an accident are typically related to equipment malfunction, component failure, or operator error. Following the terrorist events of September 11, 2001, the NRC determined that it was necessary to require certain modifications to emergency preparedness programs for operating power reactor licensees to ensure continued adequate protection of public health and safety. The agency issued these modifications to the licensees via several orders.

The NRC evaluated the emergency preparedness planning basis for nuclear power reactors given the changed threat environment. The NRC staff informed the Commission that the emergency preparedness planning basis remained valid, including scope and timing issues. The NRC staff also noted several emergency preparedness issues that required further action to better respond to the post-September 11, 2001, threat environment. As a result, the Commission directed the staff to conduct a comprehensive review of emergency preparedness regulations and guidance. The NRC staff provided the results of its review to the Commission and recommended rulemaking for enhancements to the emergency preparedness program. The Commission approved the staff's recommendation, and the rulemaking includes changes in the following areas:

- on-shift staff responsibilities
- emergency action levels for hostile action events
- emergency response organization augmentation and alternate facilities
- licensee coordination with offsite response organizations during hostile action events
- protection for onsite personnel
- challenging drills and exercises
- backup means for alert and notification systems
- emergency declaration timeliness
- emergency operations facility – performance-based approach
- evacuation time estimate updating
- amended emergency plan change process

In an effort to conduct rulemaking that is transparent and open to stakeholder participation, the NRC, in conjunction with the Federal Emergency Management Agency (FEMA), engaged stakeholders through various means during the development of this rule. The NRC and FEMA held several public meetings to discuss the proposed changes. These meetings included participants from the nuclear industry, non-governmental organizations, State and local agencies, and other interested stakeholders. The NRC also requested public comments and considered these comments in the development of the rule. The NRC and FEMA are also updating their emergency preparedness guidance documents to reflect the changes in the NRC regulations.

The new requirements should enhance the licensees' ability to prepare and implement certain emergency preparedness and protective measures in the event of a radiological emergency. These changes will also address, in part, security issues identified after the 2001 terrorist events; clarify regulations to achieve consistent emergency plan implementation among licensees; and modify certain emergency preparedness requirements to be more effective and efficient.

Status of Safety and Regulatory Issues Discussed in the Fourth U.S. National Report

Reactor Materials Degradation Issues

The reactor materials degradation issues outlined in 2007 focused on environmentally-assisted cracking of dissimilar metal welds in both PWRs and boiling water reactors (BWRs). The Wolf Creek pressurizer dissimilar metal butt weld cracking issue and the Duane Arnold jet pump riser safe end cracking event were discussed.

Wolf Creek Pressurizer Dissimilar Metal Butt Weld Cracking Issue. The discovery, in October 2006, of five circumferential indications in three dissimilar metal welds on the pressurizer at the Wolf Creek Generating Station raised safety concerns based on the size and location of the indications. This condition calls into question the degree of safety margin present in past structural integrity evaluations for dissimilar metal welds susceptible to primary water stress-corrosion cracking because of the circumferential nature of the indications and because multiple stress-corrosion cracking flaws may grow independently and ultimately grow together, significantly reducing the time from flaw initiation to leakage or rupture. To address the concern in the pressurizer surge, spray, safety, and relief nozzle welds, the NRC issued confirmatory action letters to the licensees of 40 PWR plants requesting specific inspection and leak detection enhancements. All 40 plants have completed the initial inspections, and 36 have mitigated the welds. The remaining four plants must re-inspect the remaining unmitigated welds every 4 years.

On October 22, 2008, the NRC issued Regulatory Issue Summary (RIS) 2008-25, "Regulatory Approach for Primary Water Stress Corrosion Cracking of Dissimilar Metal Butt Welds in Pressurized Water Reactor Primary Coolant System Piping." This RIS documents the current NRC regulatory approach for ensuring the integrity of primary coolant system dissimilar metal butt welds containing Alloy 182/82 in PWRs. The NRC has reviewed the industry's near-term inspection plans by monitoring the implementation of the industry's MRP-139 report, "Materials Reliability Program: Primary System Piping Butt Weld Inspection and Evaluation Guideline." The NRC is working to establish industry inspection plans for the long term. It participated with ASME to develop ASME Code Case N-770, "Alternative Examination Requirements and Acceptance Standards for Class 1 PWR Piping and Vessel Nozzle Butt Welds Fabricated with UNS N06082 or UNS W86182 Weld Filler Material With or Without Application of Listed Mitigation Activities," Dated January 26, 2009. Final incorporation of ASME Code Case N-770, with certain NRC conditions, into the *Code of Federal Regulations* is ongoing through a current rulemaking activity.

Duane Arnold Jet Pump Riser Safe End Cracking Event. Since preparation of the fourth U.S. National Report, the Boiling Water Reactor Vessels and Internals Project (BWRVIP), an industry group that provides guidance on the management of BWR materials degradation issues, evaluated the significance of the February 2007 Duane Arnold Inconel 82/182 weld cracking event to the U.S. nuclear industry. BWRVIP had previously issued guidance on the inspection of BWR welds susceptible to intergranular stress-corrosion cracking in technical report BWRVIP-75-A, "BWR Vessel and Internals Project, Technical Basis for Revisions to Generic

Letter 88-01 Inspection Schedules," dated October 2005. BWRVIP issued supplementary guidance to U.S. BWR licensees by letters dated January 23, February 28, May 24, and December 4, 2007, which requested: (1) a review of prior in-service inspection data for welds similar to those that were discovered to be cracked at Duane Arnold to verify that other indications of cracking had not been missed, and (2) expedited inspection of any welds similar to those found cracked at Duane Arnold that had not been recently examined by current, qualified inspection techniques. U.S. BWR licensees are in the process of implementing this guidance and have discovered a limited number of other indications, none of which have been of immediate safety significance. Furthermore, the BWRVIP has summarized the guidance information provided to its members in proprietary technical report BWRVIP-222, "Accelerated Inspection Program for BWRVIP-75-A Category C Dissimilar Metal Welds Containing Alloy 182," dated July 2009. With the issuance of this updated guidance, the staff considers this particular operating event to have been adequately addressed.

Unanticipated Issues Associated with Power Uprates

Potential Adverse Flow Effects. At power uprate conditions, nuclear power plants can experience significant increases in steam flow velocities. Plant experience has shown that as the higher main steamline flow passes over branch lines, it can create an acoustic resonance in the steamlines that can vary greatly from one plant to another, depending on the routing of the main steamlines and the steam dryer vintage and geometry. The acoustic resonance can create pressure waves that strike the steam dryer in BWRs with significant force. This flow could cause the stress in the steam dryer to exceed the material fatigue limits, which may result in steam dryer cracking. The acoustic resonance can also cause excessive vibration that may damage steamline and feedwater line components. For example, in 2002 and 2003, the steam dryers at Quad Cities Units 1 and 2 developed cracks and, in some cases, fractured metal parts from the steam dryer fell into the reactor pressure vessel and entered the steamlines leading to the turbine generator during EPU operation. In addition, feedwater sampling probes at Dresden Units 2 and 3 broke loose within a relatively short period of time under the higher feedwater flow conditions.

The NRC is applying lessons learned from operating experience, as well as knowledge gained from previous reviews of analyses of potential adverse flow effects, in reviewing power uprate requests for operating nuclear power plants and design certification requests for new nuclear power plants. As part of this effort, the NRC has updated relevant sections of NUREG-0800, "Standard Review Plan for the Review of Safety Analysis Reports for Nuclear Power Plants: LWR Edition," and RG 1.20, Revision 3, "Comprehensive Vibration Assessment Program for Reactor Internals during Preoperational and Initial Startup Testing," dated March 2007, to further guide NRC reviewers and the nuclear industry regarding evaluation of potential adverse flow effects.

To address the issue, BWR EPU applicants have provided complex steam dryer analyses to demonstrate the structural integrity of the steam dryers at uprated power levels. However, it has been challenging for licensees to provide acceptable steam dryer analyses, and this has significantly contributed to delays in the EPU reviews for several BWR plants. Reasons for these delays include: (1) licensees introducing new refinements to analytical methods not used in previous EPU applications, (2) the NRC identifying new issues with licensees' acoustic circuit models, (3) licensees needing to make steam dryer modifications to address analyses issues, and (4) lack of adequate plant measurement data needed for the steam dryer analyses.

To further address this issue, the industry submitted two independent topical reports to the NRC for review and approval. These reports present two independent integrated evaluation approaches and acceptance criteria for steam dryers. GE Hitachi Nuclear Energy submitted NEDC-33436P, "GEH Boiling Water Reactor Steam Dryer - Plant Based Load Evaluation," on November 7, 2008. (NEDC-33436P gives direction to refer to GE Hitachi topical report NEDC-33408P, "ESBWR Steam Dryer - Plant Based Load Evaluation Methodology," dated February 2008, which was submitted to the NRC for review and approval of similar methodology for the ESBWR.) EPRI (BWRVIP) submitted BWRVIP-194, "Methodologies for Demonstrating Steam Dryer Integrity for Power Uprate," on December 18, 2008. The NRC has begun its review of these topical reports; however, the NRC has identified the need for complementary or related topical reports, as well as additional information, to continue its review. If the NRC ultimately approves these topical reports, licensees referencing them will only need to provide the plant-specific items for review. This process should improve the review timeliness of future requests that involve evaluation of potential adverse flow effects.

PWR Post-Loss-of-Coolant-Accident Chemical Formation and PWR Sump Strainer Performance

The fourth U.S. National Report identified post-loss-of-coolant-accident chemical formation related to PWR containment sump performance as an issue. This issue remains of concern, though substantial progress has been made in resolving it. This update addresses chemical effects, but also the larger issue of sump performance, some aspects of which are not fully resolved. The NRC expects licensees to commit to specific and acceptable methods for evaluating strainer performance, as well as to make any needed plant modifications to address the results of the strainer performance evaluations. Although the NRC had planned to complete all activities related to sump strainer performance by the end of 2010, the resolution of several technical issues has been particularly challenging. Examples of the more complex issues include effects of chemical precipitate, debris generation zone of influence, and potential reactor core interactions with debris that passes through the sump strainer.

To address concerns about the potential for chemical precipitates and corrosion products to significantly block a fiber bed and increase the head loss across an emergency core cooling system sump screen, the NRC has sponsored research, issued INs, observed testing, issued review guidance, and performed detailed reviews of plant-specific evaluations. NUREG/CR-6914, Volumes 1–6, "Integrated Chemical Effects Test Project," dated December 2006, provides results from a joint NRC/U.S. nuclear industry integrated chemical effects testing program. This test program identified chemical precipitation products, and follow-up testing and analyses were performed to address the effect of chemical precipitate on head loss. Subsequent vertical loop head loss test results appear in NUREG/CR-6913, "Chemical Effects Head Loss Research in Support of Generic Safety Issue 191," dated December 2006. On the basis of these tests performed at Argonne National Laboratory, the NRC issued IN 2005-26, "Results of Chemical Effects Head Loss Tests in a Simulated PWR Sump Pool Environment," dated September 16, 2005, and IN 2005-26, Supplement 1, "Additional Results of Chemical Effects Tests in a Simulated PWR Sump Pool Environment," dated January 20, 2006.

Since the test results contained in these NUREGs demonstrated that chemical effects can be significant, the U.S. nuclear industry performed additional testing to evaluate potential chemical effects.   The NRC issued a safety evaluation report on the PWR Owners Group (Westinghouse) topical report that supports the evaluation and testing of chemical effects, WCAP-16530-NP-A, "Evaluation of Post-Accident Chemical Effects in Containment Sump Fluids to Support GSI-191," dated March 2008.   Licensees have also performed integrated head-loss testing that included chemical effects, and the NRC has visited all vendor sites that performed testing to observe tests and provide comments.   The NRC issued staff review guidance for plant-specific evaluations of chemical effects, entitled "NRC Staff Review Guidance Regarding Generic Letter 2004-02 Closure in the Area of Plant-Specific Chemical Effect Evaluations," dated March 2008.   The NRC staff is currently reviewing licensee plant-specific chemical effects evaluations, and many licensees have demonstrated an adequate evaluation of plant-specific post-loss-of-coolant-accident chemical effects.

In order to reduce the amount of debris expected to impact the sump strainer, some licensees sponsored jet impingement testing intended to show a reduced zone of influence for certain insulation and coating types.   The zone of influence determines the amount of debris generated by the postulated break and, therefore, is a significant parameter in the evaluation of the sump screen performance.   The NRC has not accepted the industry testing because of a number of concerns involving undetected flow restrictions in the test rig and the application of the test results (e.g., scaling the results to larger piping and insulation configurations in the plant).   In addition, on December 11, 2009, as a result of NRC staff questions, the test vendor identified a potential issue with the testing that may have resulted in non-conservative zones of influences in the test reports.   The NRC plans to ask licensees to demonstrate adequate strainer performance without referencing these reports, with the exception of epoxy coatings reports that can still be referenced.   The inability to take credit for the reduced zones of influence could lead to additional testing or plant modifications or both in order for affected plants to fully address the sump performance issue.

The NRC staff is currently reviewing an industry topical report by Westinghouse addressing downstream effects in the reactor vessel, WCAP-16793-NP, "Evaluation of Long Term Cooling Considering Particulate, Fibrous and Chemical Debris in the Recirculating Fluid," dated April 2009.   This document is intended to provide an acceptance criterion for licensees to use to demonstrate that debris passing through the sump strainers will not cause unacceptable impacts in the reactor core.   The NRC expects to issue a safety evaluation regarding this report in 2010. However, some questions regarding the subject matter of the report have not been fully addressed, and the industry plans additional testing to support the report.

Other Major Regulatory Accomplishments

Since its previous U.S. National Report in 2007, the NRC has issued two early site permits and one limited work authorization.   The NRC also amended its regulations concerning pressurized thermal shock, power reactor security, aircraft impact assessment, and fatigue management. The agency has also had major accomplishments in the areas of fire protection, analysis of external events, safety culture, and regulatory effectiveness.

Issuance of Early Site Permits and Limited Work Authorizations

On November 27, 2007, and August 26, 2009, the NRC issued two early site permits -- one to Dominion Nuclear North Anna, LLC, for the North Anna site in Virginia, and another to Southern Nuclear Operating Company for the Vogtle Electric Generating Plant early site permit and a limited work authorization in Georgia. The main advantage of the early site permit process is the removal of environmental contentions later in the licensing process. Successful completion of the early site permit process resolves many site-related safety and environmental issues and determines that the sites are suitable for possible future construction and operation of a nuclear power plant. The permits are valid for up to 20 years. An early site permit may be referenced in an application to the NRC for a combined license to build one or more nuclear plants on the permitted site.

Reactor Pressure Vessel Pressurized Thermal Shock

On January 4, 2010, the NRC promulgated a new regulation in 10 CFR 50.61a, "Alternate Fracture Toughness Requirements for Protection against Pressurized Thermal Shock Events." This new regulation provides an alternative set of requirements that U.S. PWR licensees may choose to implement (provided they meet certain criteria established with the regulation) to demonstrate that their facility's reactor pressure vessel will be adequately protected from failure because of a pressurized thermal shock event through the end of the facility's operating license. The NRC developed the technical basis for 10 CFR 50.61a based on a state-of-the-art probabilistic fracture mechanics methodology that accounted for, among other factors, (1) reactor pressure vessel material, (2) mechanical and chemical properties and their variability, (3) reactor pressure vessel material flaw distributions, (4) radiation damage modeling, (5) calculation of neutron fluence, (6) thermal-hydraulic modeling of pressurized thermal shock events, and (7) PRA modeling of the likelihood of a pressurized thermal shock event. It is anticipated that this new regulation may obviate the need for detailed plant-specific analyses by those licensees who would otherwise have difficulty demonstrating compliance with the NRC's original pressurized thermal shock regulation described in 10 CFR 50.61, "Fracture Toughness Requirements for Protection against Pressurized Thermal Shock Events," through the end of their facility's operating license.

Power Reactor Security

On March 27, 2009, the NRC amended its power reactor security regulations. The rulemaking: (1) makes generically applicable many of the security requirements imposed by Commission orders issued after the terrorist attacks of September 11, 2001, (2) adds several new requirements that resulted from insights gained while implementing the security orders, reviewing site security plans, and implementing the enhanced baseline inspection program and force-on-force exercises, (3) updates the regulatory framework in preparation for receiving license applications for new reactors, and (4) imposes requirements to assess and manage site activities that can adversely affect safety and security. Additionally, the NRC resolved three petitions for rulemaking as part of the effort to develop the security requirements.

This final rulemaking amended the following existing requirements within 10 CFR Part 73, "Physical Protection of Plants and Materials":

- 10 CFR 73.55, "Requirements for Physical Protection of Licensed Activities in Nuclear Power Reactors against Radiological Sabotage"
- 10 CFR 73.56, "Personnel Access Authorization Requirements for Nuclear Power Plants"
- 10 CFR Part 73, Appendix B, "Nuclear Power Reactor Training and Qualification Plan for Personnel Performing Security Program Duties"

- 10 CFR Part 73, Appendix C, "Licensee Safeguards Contingency Plans"

The amendments added two new sections to 10 CFR Part 73 and a new paragraph to 10 CFR Part 50:

- 10 CFR 73.54, "Cyber Security Requirements"
- 10 CFR 73.58, "Safety/Security Interface Requirements for Nuclear Power Reactors"
- 10 CFR 50.54(hh), "Mitigative Strategies and Response Procedures for Potential or Actual Aircraft Attacks"

There was extensive public and stakeholder participation during the development of the new requirements. The NRC extended the normal proposed rule comment period twice, offered a supplemental proposed rule comment period (related to the changes made to 10 CFR 50.54(hh)), and held meetings during the public comment period to support more informed external stakeholder feedback. The new power reactor security regulations became effective on March 31, 2010. However, due to the nuclear power plant physical changes required by the new regulations, some licensees have requested and received exemptions for the compliance date of certain elements of the rule.

Aircraft Impact Assessment

Since September 11, 2001, the issue of an airborne attack on U.S. infrastructure, including both operating and potential new nuclear power plants, has been widely discussed. The NRC has comprehensively studied the effect of an airborne attack on nuclear power plants and has undertaken a series of regulatory actions to enhance the security of nuclear power plants. Studies confirm the low likelihood that an airplane attack on a nuclear power plant would affect public health and safety, in part because of the inherent robustness of the structures. One study identified new methods plants could use to minimize damage and risk to the public in the event of any kind of large fire or explosion. Nuclear power plants subsequently implemented many of these methods, and the NRC has adopted new regulations to require both existing and new nuclear power plants to address strategies for coping with large fires or explosions from any cause, including the impact of a large, commercial aircraft.

The NRC also adopted an additional regulation for the consideration of aircraft impacts for new nuclear power reactors. This rule requires applicants for new nuclear power reactors to perform a design-specific assessment of the effects of the impact of a large, commercial aircraft, using realistic analyses. Based on the results of this assessment, applicants must identify and incorporate design features to show that the facility can withstand the effects of the aircraft impact. Applicants for all of the designs currently under NRC review have completed their aircraft impact assessments and submitted the resulting design information.

The NRC has also worked with the nuclear power industry to develop guidance for the performance of the required aircraft impact assessment. In July 2009, the NRC issued draft guidance DG-1176, "Guidance for the Assessment of Beyond-Design-Basis Aircraft Impacts," which endorsed NEI 07-13, Revision 7, "Methodology for Performing Aircraft Impact Assessments for New Plant Designs," dated May 2009. The NRC estimates that in late 2010 a RG will be issued to endorse the final version of NEI 07-13. The NRC staff will inspect the aircraft impact assessments performed by applicants for the designs currently under review.

Fatigue Management

On March 31, 2008, the NRC amended 10 CFR Part 26, "Fitness for Duty Programs," to establish enforceable requirements for the management of worker fatigue. Subpart I, of 10 CFR Part 26, "Managing Fatigue," includes new regulations that establish an integrated approach to fatigue management consisting of prevention, detection, and mitigation as the fundamental components. The rule required licensees to implement its requirements by October 1, 2009, which provided 18 months to hire and train individuals as needed to ensure proper implementation of the requirements. Subpart I strengthens the effectiveness of fitness for duty programs by ensuring that worker fatigue does not adversely affect public health and safety. In addition to the rulemaking and its associated analyses, the Commission also issued RG 5.73, "Fatigue Management for Nuclear Power Plant Personnel," in March 2009 to implement the rule.

Risk-Informed Fire Protection Infrastructure

In 2004, the NRC promulgated a rule, 10 CFR 50.48(c), which allows an operating nuclear power plant licensee to voluntarily adopt a risk-informed, performance-based fire protection program. The fire protection regulations now allow licensees to demonstrate compliance in one of two ways -- licensees may either maintain their currently approved fire protection program or transition to the risk-informed, performance-based fire protection program. The risk-informed, performance-based fire protection rule incorporates by reference the National Fire Protection Association standard 805 (NFPA 805), "Performance-Based Standard for Fire Protection for Light Water Reactor Electric Generating Plants, 2001 Edition," with several clarifications and exceptions. Licensees transitioning to 10 CFR 50.48(c) can use consensus standards on PRA quality (i.e., ASME/American Nuclear Society (ANS)-RA-Sa-2009, "Standard for Level 1/Large Early Release Frequency Probabilistic Risk Assessments for Nuclear Power Plant Applications," dated February 2009) and associated peer reviews, as endorsed in the latest revision of RG 1.200, "An Approach for Determining the Technical Adequacy of Probabilistic Risk Assessment Results for Risk-Informed Activities," published by the NRC, to help ensure the technical adequacy of their PRAs for this transition. The NRC has also sponsored research on fire protection and fire PRA issues for a number of years. One key product of this research is a joint NRC-EPRI document, NUREG/CR-6850, "EPRI/NRC-RES Fire PRA Methodology for Nuclear Power Facilities," dated September 2005.

Two nuclear stations, Oconee and Shearon Harris, volunteered to be pilot plants for the transition to a risk-informed, performance-based fire protection program, and the NRC is reviewing the pilot plants' license amendment requests. The NRC published RG 1.205, Revision 1, "Risk-Informed, Performance-Based Fire Protection for Existing Light-Water Nuclear Power Plants," in December 2009. RG 1.205, Revision 1, incorporates changes related to the following: (1) a number of lessons learned from the review of the NFPA 805 pilot plant license amendment requests and corresponding regulatory audits; (2) the issuance of NEI 04-02, Revision 2, "Guidance for Implementing a Risk-Informed, Performance-Based Fire Protection Program under 10 CFR 50.48(c)," dated April 2008; (3) interim staff guidance documents issued by the NRC; and (4) guidance needed to comply with the rule. Additional changes will be made to RG 1.205 in the future to incorporate final lessons learned from the pilot plant license amendment requests.

The NRC also issued NUREG-0800, Section 9.5.1.2, "Risk-Informed, Performance-Based Fire

Protection Program," in December 2009.   The risk-informed, performance-based approach will provide greater regulatory consistency and clarity, and provide more flexibility for licensees to address very low-risk issues without needing prior NRC staff approval.   Transitioning to this new approach includes a reassessment of the current plant fire protection program.   This could lead to the identification of previously unrecognized fire safety issues.   Subsequent resolution of these issues will result in safer plants.   To date, 51 operating reactor units, including the two pilot plants (four units), have committed to transitioning to the new rule.

Probabilistic Risk Assessment Standard for the Analysis of External Events

The NRC used a phased approach to PRA quality so that progress could be made in risk-informed activities while the necessary infrastructure (e.g., development of PRA quality standards and related industry peer review guidance) was being built.   In the initial phases, the standard for external events PRA quality was still under development, and in general, external event contributors to risk were addressed in an ad hoc fashion, including through limited or simplified quantitative analyses, qualitative arguments, and reliance on compensatory measures.

The initial consensus PRA standard for the analysis of external events for at-power operations was published in February 2009 as part of ASME/ANS RA-Sa-2009.   This standard includes internal fires, seismic events, external floods, high winds, and other external events.   In March 2009, the NRC published RG 1.200, Revision 2, which includes the NRC's endorsement (with objections and clarifications) of the PRA standard ASME/ANS RA-Sa-2009 including external events at power.   The NRC allowed a 1-year implementation period for limited-scope applications (e.g., single component technical specification changes) to enable licensees to develop or revise their PRAs, perform self-assessments and any necessary peer reviews, and address any findings of these reviews and previous reviews.   Starting in April 2010, nuclear power plant licensees who submit risk-informed licensing applications are expected to meet the guidelines in RG 1.200, Revision 2, including the external events PRA quality standard.

Future revisions to the PRA standard are expected to refine the quality expectations for internal and external event PRAs at power, as well as to incorporate additional peer review guidance and PRA standards for operations during low-power and shutdown modes.   The NRC will endorse the revisions to the standard (with objections and clarifications, as appropriate) in future revisions of RG 1.200 and will typically include a 1-year implementation period for limited-scope applications.

Regulatory Effectiveness

The NRC went through a period of expansion in which it worked aggressively to hire the highly skilled staff needed to regulate the existing fleet of operating nuclear reactors and to meet the demands for new reactor and materials license application reviews.   The agency has grown from a staff of 3,110 employees in 2004 to more than 4,000 employees today.   Although this hiring rate has decreased, the NRC is now working to meet the challenge of training and integrating a new and increasingly younger workforce, providing staff with the necessary infrastructure to successfully carry out the organization's mission.

Staffing.   The NRC recognizes that the agency must remain the employer of choice if it is to continue to be effective in accomplishing its mission.   The NRC has developed a talent acquisition plan, which includes the following elements:

- Branding – Employer branding implies name recognition and identification with a mission. The NRC is engaged in this long-term process even when not actively recruiting.

- Academic Linkages – This element includes targeted recruiting and connecting with universities. In addition to participation in career fairs, the agency has University Champions who facilitate relationships between the NRC and individual universities to aid in recruiting, and it engages faculty and administration in the agency's work through grants and scholarships.

- Mission Driven – NRC staff members want to know how their work relates to the agency mission and how they are making a difference. This is, perhaps, the NRC's most important recruitment and retention tool.

Responses to employee viewpoint surveys show that NRC efforts to hire and retain a highly motivated workforce are working. In 2007 and 2009, the NRC was ranked as the best place to work in the Federal Government. The results of the 2009 survey reflect that employees feel strongly engaged, understand how their work contributes to the agency's mission, and view their work as meaningful and important. Survey results also indicate that employees agree that they have the training, development, information, and skills needed to perform their work.

Training. Nearly half of all NRC staff members have been with the agency for less than 4-years. Rapidly training and integrating this large number of new employees into the agency is a significant challenge. The NRC uses an integrated approach to learning to provide new employees with consistent information from branch to branch and division to division.

For example, the agency has adopted an enterprise-wide leadership development program for all workforce segments, from entry-level through the Senior Executive Service. The program focuses on development of 28 defined Federal Government-wide leadership competencies. To assist new employees, the NRC is implementing a virtual orientation center. This advanced training tool allows new hires to enter a computer-generated or virtual world where they can obtain information about the NRC organization, mission, and employee benefits before starting their first day of work.

Additionally, the NRC offers position-specific training to accompany this generic orientation. The main NRC offices, such as the Office of Nuclear Reactor Regulation, have developed a qualification program that consists of three parts: general requirements, position-specific requirements, and oral qualification boards. The NRC is continuing to develop its qualification plans and other position-specific training, such as for project engineers and project managers. It is also identifying course needs at its Technical Training Center and Professional Development Center.

Knowledge Management. The NRC has incorporated knowledge management into its strategic workforce planning. The goal is to identify short- and long-term critical skill gaps to enable the agency to anticipate change. To this end, the NRC attempts to spot workforce trends and projections and to close anticipated skill gaps through both training and development and knowledge management.

The NRC uses an agency-wide knowledge management plan that serves as a framework to integrate new and existing approaches that generate, capture, and transfer knowledge and information relevant to the NRC's mission. The following are some of the near-term and long-term strategies for this plan:

- capture relevant critical knowledge from departing personnel
- recapture departed knowledge where possible
- communicate leadership's expectation for a knowledge-sharing culture
- formalize knowledge management values and principles
- incorporate knowledge management within process workflows

Some of the knowledge management and transfer activities used to accomplish these goals include the following:

- Branch Chief and Team Leader Seminars - As a community of practice, the branch chiefs and team leaders meet monthly and hear presentations by agency experts on topics such as performance management, budget, and communications.

- Video Interviews – The NRC conducted a pilot project to capture knowledge from retiring senior staff using video interviews. The interviews included questions about licensing issues, recruiting and mentoring new hires, leadership, operations center experience, and reactor licensing performance metrics.

- Web Sites – The NRC has developed the NRC Knowledge Center Web page that links a number of communities and topics. This page is supplemented by office-specific knowledge management programs.

Finally, the NRC makes prudent, targeted use of retention incentives and pension offset waivers (rehiring annuitants without reduction of salary or pension) in order to retain highly qualified employees and as a knowledge management tool. Such incentives are particularly useful for unusual occupations or highly specialized disciplines for which candidates may be scarce.

Section 8.1.5.2 of this report discusses the human resources in more detail.

Safety Culture Initiatives

Based on lessons learned from the Davis-Besse reactor pressure vessel head degradation event and other considerations, the NRC enhanced the Reactor Oversight Process to more fully address safety culture and identify safety culture problems earlier so that corrective steps can be taken to address the problems and prevent further plant performance degradation.

In July 2006, the NRC implemented revisions to the Reactor Oversight Process inspection and assessment processes related to safety culture. In 2008, the NRC conducted a self-assessment to review the changes to the Reactor Oversight Process over the initial 18-month implementation period. Lessons learned from the initial 18-month implementation period and from the Palo Verde supplemental inspection resulted in changes to inspection procedures and program guidance.

In November 2009, the agency published a draft Safety Culture Policy Statement in the *Federal Register* that set forth the expectation that all licensees and certificate holders establish and maintain a positive safety culture. Similarly, given the NRC's safety and security mission, the NRC recognizes the importance of maintaining its own strong safety culture and the need to continuously seek to improve its internal organizational effectiveness.

The agency is implementing several initiatives to improve safety culture. Also, the agency uses the periodic Safety Culture and Climate Survey by the Office of the Inspector General as a means to assess the effectiveness of these new and existing safety culture efforts. The latest survey took place in 2009, and the NRC is addressing the survey responses to maintain areas identified as strengths and to improve areas identified as challenges.

Section 10.4 of this report discusses safety culture in more detail.

The NRC's Main Challenges

The NRC identified major challenges for the future in its Strategic Plan for 2008-2013, dated February 2008. Challenges, summarized below and detailed in Appendix A to this report, arise from the changing regulatory environment and external factors.

- The NRC expects to receive additional applications from entities that want to build and operate new nuclear power plants.

- Increasing quantities of spent nuclear fuel will be held in interim storage at reactor sites or transported to centralized interim storage sites awaiting permanent disposal.

- The NRC will continue to coordinate with a wide array of Federal, State, local, and Tribal authorities on issues related to license renewal, new reactor licensing, homeland security, emergency planning, and environmental protection.

The NRC recognizes that these changes will create an even greater need for effective and open communication with public stakeholders about a variety of issues. These include the safety and security of existing and proposed nuclear power plants and other licensed facilities and materials, emergency preparedness, and the impact on public health and safety and the environment from medical, academic, and industrial uses of licensed materials.

The following key external factors could cause challenges:

- receipt of new reactor license applications
- a significant operating incident (domestic or international)
- a significant terrorist incident
- timing of the U.S. Department of Energy (DOE) application for the high-level waste repository at Yucca Mountain and related activities[2]
- legislative initiatives

---

[2] In March 2010, DOE filed a motion to withdraw its application from NRC review. Section 19.8 of this report discusses radioactive waste in more detail.

NRC Major Management Challenges

By law, the Inspector General of each Federal agency (as discussed in Article 8) identifies the agency's most serious management and performance challenges facing the agency and assesses progress in addressing them.   The NRC's Inspector General's annual assessment of the major management challenges confronting the agency appear on the NRC's public Web site. The 2009 assessment report described the following main challenges, given in more detail in Appendix B to this report.

- protection of nuclear material used for civilian purposes
- managing information to balance security with openness and accountability
- ability to modify regulatory processes to meet a changing environment to include the licensing of new nuclear facilities.
- oversight of radiological waste
- implementation of information technology and information security measures
- administration of all aspects of financial management
- managing human capital

PART 2

ARTICLE 6.   EXISTING NUCLEAR INSTALLATIONS

Each Contracting Party shall take the appropriate steps to ensure that the safety of nuclear installations existing at the time the Convention enters into force for that Contracting Party is reviewed as soon as possible.   When necessary in the context of this Convention, the Contracting Party shall ensure that all reasonably practicable improvements are made as a matter of urgency to upgrade the safety of the nuclear installation.   If such upgrading cannot be achieved, plans should be implemented to shut down the nuclear installation as soon as practically possible.   The timing of the shutdown may take into account the whole energy context and possible alternatives, as well as the social, environmental, and economic impact.

This section explains how the United States ensures the safety of nuclear installations in accordance with the obligations in Article 6.   It covers the reactor licensing and major oversight processes in the United States.   This section also discusses programs for rulemaking, fire protection regulation, decommissioning, research, and programs for public participation. The U.S. Nuclear Regulatory Commission (hereafter referred to as the NRC, Commission, agency, or staff) posts the major results of assessments on the agency's public Web site at http://www.nrc.gov.   This update includes expectations about early site permits and design certification applications, current experience, and revised details about programs.

6.1   Introduction

The mission of the NRC is to license and regulate the Nation's civilian use of byproduct, source, and special nuclear materials in order to protect public health and safety, promote the common defense and security, and protect the environment.   The NRC's primary goal is safety.   The agency achieves this goal by ensuring that licensee performance is at or above acceptable safety levels.   The NRC's licensees are responsible for designing, constructing, and operating nuclear facilities safely, while the NRC is responsible for the regulatory oversight of the licensees. Five strategic outcomes for this goal are specified:

(1)   No nuclear reactor accidents.
(2)   No inadvertent criticality events.
(3)   No acute radiation exposures resulting in fatalities.
(4)   No releases of radioactive materials that result in significant radiation exposures.
(5)   No releases of radioactive materials that cause significant adverse environmental impacts.

The NRC met all of its safety strategic outcomes in fiscal years (FYs) 2008 and 2009.

The NRC also uses performance measures to determine whether the agency has met its safety goal.   The NRC met its performance measures in FYs 2008 and 2009.   Currently the NRC uses six performance measures.

The first measure analyzes nuclear power plant performance based on a large number of performance indicators and inspection findings.

The second measure tracks significant precursor events at nuclear power plants determined by the likelihood of an event adversely impacting safety.

The third performance measure indicates whether the NRC identifies significant issues in a nuclear power plant during inspections conducted under the Reactor Oversight Process.

The fourth measure tracks the trends of several key indicators of nuclear power plant safety. This measure is the broadest measure of the safety of nuclear power plants, incorporating the performance results from all plants to determine industry average results.

These four measures indicated that the nuclear power plants were safely operated, and that the events that did occur were of relatively minor significance.

The other two measures address harmful radiation exposures to the public and occupational workers and radiation exposures that harm the environment. Neither of these measures exceeded their targets in FY 2009.

6.2 Nuclear Installations in the United States

Annex 1 to Part 2 of this report lists all 104 nuclear installations in the U.S., as discussed in NUREG-1350, Volume 21, "NRC Information Digest 2009-2010," dated August 2009, available on the agency's Web site.

6.3 Regulatory Processes and Programs

6.3.1 Reactor Licensing

To construct and operate a nuclear reactor, an entity must submit an application to the NRC for a safety and environmental review. The public has opportunities to participate through a hearing process. The NRC licensed all current operating nuclear plants under the detailed two-step process specified in Title 10 of the *Code of Federal Regulations* (10 CFR) Part 50, "Domestic Licensing of Production and Utilization Facilities," first issuing a construction permit and then an operating license. Since 1976, the NRC has not received applications to construct a new reactor under 10 CFR Part 50. A new single-step process specified in 10 CFR Part 52, "Licenses, Certifications, and Approvals for Nuclear Power Plants," provides direction for issuing a combined license for construction and operation of a new reactor. The NRC has received 18 combined license applications for 28 reactors. In addition, largely because of the favorable incentives created by the U.S. Congress in the Energy Policy Act of 2005, the industry has submitted applications for three additional design certifications, one design certification amendment, and one design certification request to amend the rule for aircraft impact. To date, the NRC has issued four early site permits and one limited work authorization in August 2009. To date, the agency has not issued any combined licenses. Article 18 provides more detail about the 10 CFR Part 52 regulations.

The NRC's reactor licensing process provides for the review and approval of changes after initial licensing. The process allows amendments to the operating license to support plant changes, license renewal, changes of ownership and license transfer, exemptions and relief from NRC regulations, and increases in the reactor power level (i.e., power uprates). This report provides additional discussion of the process in the introduction and other articles (i.e., Articles 14, 17 and 18).

44

6.3.2 Reactor Oversight Process

Through its Reactor Oversight Process, the NRC continuously oversees nuclear power plants to verify that they are being operated in accordance with the agency's rules and regulations.   The NRC has full authority to take whatever action is necessary to protect public health and safety, and may demand immediate licensee actions, up to and including a plant shutdown.

The Reactor Oversight Process uses both inspection findings and performance indicators to assess the performance of each plant within a regulatory framework of seven cornerstones of safety.   Toward that end, the NRC has at least two resident inspectors stationed at each plant who perform a program of baseline inspections.   To supplement these continuous inspections, regional inspection specialists conduct periodic inspections of each plant in his or her region.   If a particular plant exceeds established thresholds during these routine inspections, the regional inspectors may perform special inspections and take additional actions to ensure that the plant addresses these significant issues.   The NRC communicates the results of its oversight process by posting plant-specific inspection findings and performance indicator information on the agency's public Web site.   The NRC also conducts public meetings with licensees to discuss the results of its assessments of licensee performance.

The NRC assesses the Reactor Oversight Process annually and evaluates its overall effectiveness through the agency's success in meeting its pre-established goals (i.e., performance metrics) and intended outcomes.   The NRC issued its latest report on the subject, SECY-10-0042, "Reactor Oversight Process Self-Assessment for Calendar Year 2009," on April 7, 2010.

The results of the calendar year 2009 self-assessment indicated that the Reactor Oversight Process met its program goals and achieved its intended outcomes.   The staff found the Reactor Oversight Process to be objective, risk informed, understandable, and predictable, and it met the agency's strategic goals of ensuring safety and security.   The staff implemented several Reactor Oversight Process improvements in calendar year 2009 to address issues raised by the Commission, and obtained from internal and external stakeholder feedback.

The staff continues to improve the performance indicators and explore potential new indicators to ensure that the performance indicators program provides meaningful input to the Reactor Oversight Process.   The inspection program independently verified that plants were operated safely and securely, and ensured that sites remained staffed with knowledgeable and experienced inspectors.   The significance determination process remained an effective tool for determining the safety and security significance of identified performance issues in a timely manner.   The self-assessment provided for regulatory oversight in identifying licensee performance issues and determining appropriate regulatory response.   The staff continues to solicit input from the NRC's internal and external stakeholders to further improve the Reactor Oversight Process based on stakeholder feedback and lessons learned.

Based on its calendar year 2009 self-assessment, the NRC plans the following significant actions or ongoing activities in 2010 to improve the efficiency and effectiveness of the Reactor Oversight Process:

- Develop a framework for evaluating the efficacy of potential new performance indicators for use in the Reactor Oversight Process.

- Continue to emphasize the availability and use of operating experience in the inspection program and further integrate this emphasis into the inspection guidance.

- Conduct additional training on the significance determination process based on input from the partnering initiative, which provided valuable insights regarding areas where training was lacking or can be improved.

- Report to the Commission on how the proposed enhancements to the force-on-force physical protection significance determination process would alter the 2009 exercise findings.

- Revise program guidance, as necessary, to align with the Commission's safety culture policy statement, once it has been completed.

6.3.3 Industry Trends Program

The NRC staff implemented the Industry Trends Program in 2001. The agency continues to develop the program as a means to confirm that the nuclear power industry is maintaining the safety of operating power plants and to increase public confidence in the effectiveness of the NRC's processes. The agency uses industry-level indicators to identify adverse trends in performance. After assessing industry trends for safety significance, the NRC responds as necessary to any identified safety issues, including adjusting the inspection and licensing programs if necessary. One important output of the Industry Trends Program is the annual agency performance measures reported to Congress on the number of statistically significant adverse industry trends. The NRC Performance and Accountability Report includes this outcome measure.

In addition to long-term trending of the data to identify statistically significant adverse trends, the NRC staff uses a statistical approach based on prediction limits to identify potential short-term, year-to-year emergent issues before they become long-term trends. Short-term fiscal year (FY) 2009 data did not identify any issues that warranted additional analysis or significant adjustments to the nuclear reactor safety inspection or licensing programs.

The Reactor Oversight Process uses both plant-level performance indicators and inspections to provide plant-specific oversight of safety performance, whereas the Industry Trends Program provides a means to assess overall industry performance using industry-level indicators. The NRC evaluates issues that are identified through either program using information from agency databases and addresses those determined to have generic safety significance, including generic safety inspections under the Reactor Oversight Process, the generic communications process, and the generic safety issue process.

Based on the information currently available from the industry-level indicators and the Accident Sequence Precursor Program (discussed in Section 6.3.4), no statistically significant adverse industry trends have been identified through FY 2009.

The staff has recently implemented the Baseline Risk Index for Initiating Events (BRIIE), a new indicator that monitors nine risk-significant initiating events for boiling-water reactors (BWRs) and 10 events for pressurized-water reactors (PWRs) (the additional event category is steam generator tube rupture).

The BRIIE concept provides a two-level approach to industry performance monitoring: (1) it tracks several types of events that could potentially start ("initiate") a challenge to a plant's safety systems; (2) it assigns a value to each initiating event according to its relative importance to the plant's overall risk of damage to the reactor core, then calculates an overall indicator of industry safety performance.

The first level (referred to as Tier 1 performance monitoring) tracks and counts the number of times the initiating events that have an impact on plant safety occur in nuclear power plants during the year. The number of times that each event occurs is compared with a predetermined number of occurrences for that event. If the predetermined number is exceeded, one can infer possible degradation of industry safety performance. This annual tracking allows the NRC to intervene and engage the nuclear industry before any long-term adverse trends in performance emerge.

The second level (referred to as Tier 2 performance monitoring) addresses the risk to plant safety and core damage that each of the initiating events contributes. Each of the events is assigned an importance value, a ranking according to its relative contribution to overall risk to plant safety. The greater the contribution of the event to overall risk, the higher the importance value that is assigned to the event. Using statistical methods, the importance values are combined with the number of times the events occur during the year to calculate a number that indicates how much the overall industry risk of damage to the reactor core has changed from a baseline value. The NRC Performance and Accountability Report notes if this combined industry value reaches or exceeds a threshold value of 1×10^{-5} per reactor critical year, along with actions that have already been taken or are planned in response.

None of the initiating events tracked in Tier 1 exceeded its prediction limit in FY 2009. The BRIIE Tier 2 combined industry value in FY 2009 (i.e., -2.36×10^{-6} per reactor critical year) indicates better than baseline industry performance and is well below the established reporting threshold of 1×10^{-5} per reactor critical year.

SECY-10-0028, "FY 2009 Results of the Industry Trends Program for Operating Power Reactors and Status of Ongoing Development," dated March 16, 2010, and available on the NRC public Web site, provides more details.

6.3.4 Accident Sequence Precursor Program

The Accident Sequence Precursor Program systematically evaluates U.S. nuclear power plant operating experience to identify, document, and rank the operating events that are most likely to lead to inadequate core cooling and severe core damage (i.e., precursors).

To identify potential precursors, the NRC reviews plant events from licensee event reports and inspection reports. The staff then analyzes any identified potential precursors by calculating the probability of an event leading to a core damage state. A plant event can be one of two types, either (1) an occurrence of an initiating event, such as a reactor shutdown or a loss of offsite power, with or without any subsequent equipment unavailability or degradation, or (2) a degraded plant condition, depicted by the unavailability or degradation of equipment without the occurrence of an initiating event.

The Accident Sequence Precursor Program considers an event with a conditional core damage probability or an increase in core damage probability greater than or equal to 1×10^{-6} to be a precursor. The Accident Sequence Precursor Program defines a *significant* precursor as an event with a conditional core damage probability or an increase in core damage probability greater than or equal to 1×10^{-3}.

The Accident Sequence Precursor Program has the following objectives:

- Provide a comprehensive, risk-based view of nuclear power plant operating experience and a measure for trending nuclear power plant core damage risk.

- Provide a partial check on dominant core damage scenarios predicted by probabilistic risk assessments (PRAs).

- Provide feedback for regulatory activities.

The NRC also uses the Accident Sequence Precursor Program to monitor performance against the safety goal established in the agency's Strategic Plan. Specifically, the program provides input to the following performance measures:

- Zero events per year identified as a *significant* precursor of a nuclear reactor accident.

- No more than one significant adverse trend in industry safety performance (determination principally made from the Industry Trends Program but partially supported by Accident Sequence Precursor results).

The staff completed precursor trend analyses as part of the annual Accident Sequence Precursor Program status report provided to the Commission in SECY-09-0143, "Status of the Accident Sequence Precursor Program and the Standardized Plant Analysis Risk Models," dated September 29, 2009. The report provided insights such as the following:

- No *significant* precursors were identified in FY 2009. The last *significant* precursor was identified in FY 2002 (i.e., multiple degraded conditions at Davis-Besse).

- A statistically significant decreasing trend was detected in the occurrence rate of all precursors during the FY 2001–2008 period.

- During the same period, statistically significant decreasing trends were detected for three groups of precursors: (1) precursors with a conditional core damage probability or an increase in core damage probability greater than or equal to 10^{-4}, (2) precursors involving degraded conditions, and (3) precursors that occurred at PWRs.

6.3.5 Operating Experience Program

The NRC launched the revised Operating Experience Program in January 2005, recognizing that the effective use of operating experience is important for the agency's safety mission. Under the current NRC Strategic Plan, the agency is committed to "evaluate domestic and international operating events and trends for risk significance and generic applicability in order to improve NRC programs" as part of its effort to achieve the goal of safety. As a result, the NRC's emphasis on the effective use of operating experience remains strong.

The fundamental aim of the Operating Experience Program is to collect, evaluate, communicate, and apply operating experience information to achieve the NRC's principal safety mission of protecting people and the environment. Operating experience is reported to the NRC identified in licensee event notifications and in many other reports that are submitted under licensee reporting requirements, and described in reports of operating experience at foreign facilities. Sources of foreign operating experience include International Nuclear Event Scale events and Incident Reporting System reports. NRC staff systematically screens nuclear reactor-related operating experience for safety significance and generic implications. The NRC staff also determines the need for further action and application of lessons learned related to plant operating experience.

To support its safety mission, the NRC increased resources dedicated to the review of operating experience and instituted a clearinghouse. The clearinghouse collects, stores, screens, and communicates operating experience; conducts and coordinates the evaluation of operating experience; tracks the application of operating experience lessons learned; and coordinates NRC operating experience activities with other organizations performing related functions.

Upon launching the program, the NRC developed an internal Web site to provide a centralized source for accessing reactor operating experience information. This Web site is a gateway to the agency's operating experience document collections, contacts, search tools, sources, and reference material. In addition, the NRC created an operating experience community forum to quickly disseminate operating experience to the appropriate technical staff. The agency's public Web site contains all of the NRC's event-related reports.

Section 19.7 of this report provides more information about this program.

6.3.6 Generic Issues Program

The U.S. Congress mandated the NRC's agencywide Generic Issues Program to address issues that have significant generic implications related to safety or security that cannot be more appropriately addressed by other regulatory programs or processes. Sources of candidate generic issues include safety evaluations, operational events, and suggestions from NRC staff members, outside organizations, or members of the general public. Other existing programs generally address emergent issues that demand immediate attention (e.g., issues that may require plant shutdown) so that quick decisions can be made. The NRC maintains a complete list of all generic issues in NUREG-0933, "Resolution of Generic Safety Issues," published most recently in August 2008.

In order to efficiently use program resources and promote timeliness, the following seven criteria describe those issues that are appropriate for processing through the program:

(1)   affects public health and safety, security, or the environment (this includes a risk threshold)
(2)   applies to two or more facilities
(3)   cannot be readily addressed through other regulatory processes or voluntary industry initiatives
(4)   can be resolved by new or revised regulation, policy, or guidance
(5)   risk or safety significance can be adequately determined or estimated
(6)   well defined and discrete
(7)   may involve review, analysis, or action by the licensee

Recent major program changes are intended to: (1) ensure timeliness of issue resolution, (2) clarify roles and responsibilities of the participating offices, (3) increase participation of the nuclear industry stakeholders and other stakeholders, as appropriate, and (4) establish clear interfaces between the Generic Issues Program and other NRC processes and activities that are used to address generic issues outside the Generic Issues Program.

6.3.7 Rulemaking

The NRC's regulations, also called rules, impose requirements that licensees must meet to obtain or retain a license or certificate to use nuclear materials or to operate a nuclear facility. The technical staff usually proposes a rule or a change to a rule because of a perceived need to protect public health and safety. However, any member of the public may petition the NRC to develop, change, or rescind a rule. The impetus for a proposed rule could be a requirement issued by the Commission, a petition for rulemaking submitted by a member of the public, or research results that indicate a need for a rule change. The NRC publishes the proposed rule in the *Federal Register* for public comment. Once the public comment period has closed, the staff analyzes the comments, makes any needed changes, and forwards the final rule for approval, signature, and publication in the *Federal Register*.

The NRC uses http://www.regulations.gov to provide an easy means for members of the public to access and comment on NRC rulemaking actions. Accessible through the Internet, Regulations.gov contains proposed rulemakings that have been published in the *Federal Register*, petitions for rulemaking, and other types of documents related to rulemaking proceedings.

The Commission must approve each final rule that involves significant matters of policy. The Executive Director for Operations is authorized to approve final rules that do not involve policy changes. Once approved, the final rule is published in the *Federal Register* and will become effective 30 days or after from the date of publication. The section of this report on major regulatory accomplishments summarizes the significant nuclear reactor-related rules issued since the previous U.S. National Report.

6.3.8 Fire Regulation Program

The NRC has three main foci in fire protection regulation: (1) implementation of the new risk-informed, performance-based fire protection licensing basis (10 CFR 50.48(c)); (2) resolution of the fire-induced multiple spurious operation/circuit analysis issue; and (3) resolution of licensees' non-conforming post-fire operator manual actions. To support the implementation of 10 CFR 50.48(c), the NRC issued Regulatory Guide (RG) 1.205 "Risk-Informed, Performance-Based Fire Protection for Existing Light-Water Nuclear Power Plants," in May 2006, and RG 1.205, Revision 1, in December 2009. As of March 2010, approximately 50 reactor units are committed to transitioning to 10 CFR 50.48(c). Two nuclear stations, Oconee and Shearon Harris, volunteered as pilot plants for the transition, and the NRC is reviewing their license amendment requests. The NRC is also developing guidance to conduct triennial fire inspections of plants after they complete their transitions to the 10 CFR 50.48(c) licensing bases. The Survey of Current Regulatory and Safety Issues section of this report provides further information on the NRC's risk-informed fire protection infrastructure.

For plants that are not transitioning to the risk-informed, performance-based fire protection rule, the NRC published RG1.189, Revision 2, "Fire Protection for Nuclear Power Plants," in November 2009. Revision 2 of RG 1.189 offers clear guidance on acceptable means for addressing multiple spurious operations, as well as general fire protection guidance. Where appropriate, it endorses approaches that industry is developing to assist licensees in their efforts to meet regulatory requirements as, provided in Nuclear Energy Institute (NEI) guidance document NEI 00-01, Revision 2, "Guidance for Post-Fire Safe-Shutdown Circuit Analysis," dated May 2009.

The NRC continues to engage with stakeholders to develop an acceptable method to resolve the issue of circuit analysis actuations. Enforcement Guidance Memorandum (EGM)-09-002, "Enforcement Discretion for Fire Induced Circuit Faults," dated May 2009, gives licensees until May 2010 to identify non-compliances, implement compensatory measures, and enter non-compliances into their corrective action program while under enforcement discretion. Licensees have until November 2012 to implement the resolutions.

On the topic of operator manual actions, the NRC issued Regulatory Issue Summary (RIS) 2006-10, "Regulatory Expectations with Appendix R Paragraph III.G.2 Operator Manual Actions," dated June 30, 2006, and developed NUREG-1852, "Demonstrating the Feasibility and Reliability of Operator Manual Actions in Response to Fire," dated October 2007. These documents were issued as a follow-on to the withdrawn proposal to amend 10 CFR 50, Appendix R, Section III.G.2, "Fire Protection Program for Nuclear Power Facilities Operating Prior to January 1, 1979," via a rulemaking that would have codified the use of operator manual actions meeting specified criteria. Via EGM 2007-004, "Enforcement Discretion for Post-Fire Manual Actions Used as Compensatory Measures for Fire Induced Circuit Failures," the NRC granted enforcement discretion to licensees to resolve issues involving postfire operator manual actions by March 2009. Those licensees seeking exemptions to the regulations or changes to their approved fire protection program for the use of operator manual actions have submitted their requests. Currently, the NRC staff is reviewing 11 such requests.

The NRC has an active fire research program that develops the technical bases for ongoing and future regulatory activities in fire protection and fire risk analysis. The NRC's current research program includes the following activities:

- developing and improving fire risk analysis methods and tools
- applying these methods and tools to develop risk insights
- collecting, generating and analyzing fire related data
- verifying, validating and improving fire models for regulatory use
- performing specialized fire testing on electrical cables for both hot shorts and fire properties
- evaluating shipping casks for beyond design basis fire conditions
- evaluating methods to predict operator performance during fire conditions
- providing specialized training on the fire PRA and human reliability analysis methods and performing fire modeling
- knowledge management

The fire research program supports the agency's strategic goals of safety and effectiveness and partners with other organizations with similar missions such as the National Institute of Standards and Technology and the Electric Power Research Institute (EPRI), the University of Maryland, and international groups such as the Organisation for Economic Co-operation and Development Committee on the Safety of Nuclear Installations.

6.3.9 Decommissioning

The decommissioning process consists of a series of integrated activities, beginning with the nuclear facility transitioning from "active" to "decommissioning" status and concluding with termination of the license, and release of the site. The NRC has adopted extensive regulations to ensure that decommissioning is accomplished safely and that residual radioactivity is reduced to a level that permits release of the property for unrestricted use (10 CFR 20 Subpart E, "Radiological Criteria for License Determination"). The NRC reviews and approves license termination plans, conducts inspections, processes license amendments, and monitors the status of activities to ensure that radioactive contamination is reduced or stabilized. In addition, the decommissioning process includes several opportunities for public involvement.

The design criteria for new facility construction at 10 CFR 20.1406, "Minimization of Contamination," require applicants to describe how facility design and procedures will facilitate eventual decommissioning and minimize, to the extent practicable, the generation of radioactive waste. Furthermore, the safety standards on decommissioning promulgated by the International Atomic Energy Agency (IAEA) include considerations, which the United States supports, for future decommissioning provisions in the conceptual design of nuclear facilities.

NRC regulations and guidance (e.g., NUREG-1577, Revision 1, "Standard Review Plan on Power Reactor Licensee Financial Qualifications and Decommissioning Funding Assurance," dated February 1999) describe requirements and processes to review power reactor licensee financial qualifications and methods of providing decommissioning funding assurance.

Spent fuel can remain stored in the spent fuel pools or in dry cask storage facilities until a geologic repository is built and operating. The NRC regulations in 10 CFR Part 50 and 10 CFR Part 72, "Licensing Requirements for the Independent Storage of Spent Nuclear Fuel, High-Level Radioactive Waste, Reactor-Related Greater Than Class C Waste," contain licensing requirements to maintain spent fuel integrity. The Commission, in issuing its Waste Confidence Decision in 1990, found that spent fuel can be stored safely in spent fuel pools or in onsite independent spent fuel storage installations without significant environmental impacts for at least 30 years beyond the plant's licensed life (which may include the term of a renewed license).

6.3.10 Reactor Safety Research Program

The NRC conducts reactor safety research to support its mission of ensuring that its licensees safely design, construct, and operate nuclear reactor facilities. The agency carries out this research program to (1) identify, evaluate, and resolve safety issues, (2) ensure that an independent technical basis exists to review licensee submittals, (3) evaluate operating experience and results of risk assessments for safety implications, and (4) support the development and use of risk-informed regulatory approaches. In conducting the Reactor Safety Research Program, the NRC anticipates challenges posed by the introduction of new technologies. The NRC continues to seek out opportunities to leverage its resources through both domestic and international cooperative programs and to provide enhanced opportunities for stakeholder involvement and feedback on its research program.

The NRC conducts pre-application reviews for advanced non-light-water reactor designs under the Reactor Safety Research Program. In the pre-application phase, the NRC interacts with prospective design certification applicants to address topics that would benefit both the applicant and the staff in preparing for a design certification application. The Commission's Policy Statement on Advanced Reactors (SECY-93-087, "Policy, Technical, and Licensing Issues Pertaining to Evolutionary and Advanced Light-Water Reactor Designs," dated April 2, 1993) encourages early interactions on such advanced designs so as to facilitate the resolution of safety issues early in the design process. In addition, the agency will conduct research to address technical issues that it anticipates will arise during its review of advanced reactor designs.

6.3.11 Special Programs for Public Participation

The NRC views nuclear regulation as the public's business and, as such, believes it should be transacted as openly and candidly as possible to maintain and enhance the public's confidence. Ensuring appropriate openness explicitly recognizes that the public must be informed about, and have a reasonable opportunity to participate meaningfully in, the NRC's regulatory processes.

The NRC extends opportunities to participate in the agency's regulatory process to a diverse body of stakeholders, including the general public; Congress; other Federal, State, and local governments; Indian Tribes; industry; technical societies; the international community; and citizen groups.   Numerous NRC programs and processes provide the public with accessibility to NRC staff and resources; seek to make communication with stakeholders more clear, accurate, reliable, objective, and timely; and help to ensure that the reporting of nuclear power plants performance is open and objective.   The agency has developed Web sites to disseminate timely, accurate information regarding issues of interest to the public or events at nuclear facilities. The NRC elicits public involvement early in the regulatory process so as to address any safety concerns in a timely manner.   In addition to the formal petition and hearing processes integrated into the licensing program, the agency also uses feedback forms at public meetings to obtain public input.   Section 7.2.2 of this report provides more information about the NRC's hearing process.

The NRC manages its rulemaking dockets using the Federal Docket Management System, a tool that provides a single point of access at http://www.regulations.gov across the Federal Government.   The public can now access more than 7,400 NRC documents related to almost 300 rulemaking actions conducted by the NRC from January 1999 through March 2008.   As agency viewers of the Federal Docket Management System, NRC employees are able to access these documents, including public comments, petitions for rulemaking, *Federal Register* notices, and their supporting materials.

Fostering an environment in which safety issues can be openly identified without fear of retribution is of paramount interest to the NRC.   The agency has established tools for the public, industry, and NRC employees to use to raise safety concerns, including the petition process under 10 CFR 2.206, "Requests for Actions under This Subpart," the safety-conscious work environment policy, and the allegation program.

The NRC petition process regulations in 10 CFR 2.206 allow any member of the public to raise potential health and safety concerns and ask the agency to take specific enforcement actions against a licensee.   If warranted, the NRC can modify, suspend, or revoke a license, or take other appropriate enforcement action, to resolve a problem identified in the petition.   Recent changes made to the petition process emphasize a timely response to the petitioner and encourage increased, direct involvement of the petitioner (in addition to involvement of the licensee) by allowing the petitioner to personally address the petition review board and comment on the agency's decision.

Any member of the public may petition the NRC to develop, change, or rescind a rule under 10 CFR 2.802, "Petition for Rulemaking."   Upon receiving the petition, the NRC publishes it in the *Federal Register* for public comment.   The NRC staff will evaluate the petition and any comments received and may either grant or deny the petition or, in some instances, partially consider or deny the petition.   In considering the petition, the NRC will publish a proposed rule that would address the concern included in the petition.   This action would be followed by a period for public comments and the publication of a final rule.   In denying a petition, the NRC publishes a notice of denial in the *Federal Register*.   This notice of denial will address any public comments received and the reason for denying the petition.

The NRC encourages workers in the nuclear industry to take their concerns directly to their employers and is vigilant about fostering a safety-conscious work environment that encourages such reporting.   The NRC expects licensees and other employers subject to NRC authority to establish and maintain a work environment where employees do not fear retribution by a licensee for raising concerns about safety or regulatory issues.   Additionally, workers and members of the public may bring their concerns relating to safety or regulatory issues directly to the NRC.   The agency established a toll-free safety hotline for reporting such concerns, and NRC management, staff, and inspectors, including the resident inspectors at plant sites, are trained and available to receive such concerns.

Historically, industry workers or members of the public report approximately 600 potential allegations directly to the NRC allegation program each year.   The NRC developed the allegation program to establish a formal process for evaluating and responding to each issue. The primary purpose of the program is to provide an alternative method for individuals to raise safety or regulatory issues and to have them addressed.   About 60 percent of the issues that are reported to the NRC are from licensee employees, employees of contractors to licensees, or former employees of licensees or contractors.   Given sufficient information, the staff will evaluate each issue to determine whether it can verify the issue and, if so, the effect of the issue on plant safety.   The evaluation either involves an engineering review, inspection, or investigation by NRC staff or an evaluation by the licensee that is independently assessed by the NRC staff.   Historically, the NRC has been able to substantiate 30 percent of the allegations received.   If the evaluation reveals a violation of regulatory requirements, the agency takes appropriate enforcement action.   Additionally, the NRC informs in writing the individual who raised the issue of the results of its evaluation, except in limited instances when sensitive security-related matters are discussed

ARTICLE 7. LEGISLATIVE AND REGULATORY FRAMEWORK

1. **Each Contracting Party shall establish and maintain a legislative and regulatory framework to govern the safety of nuclear installations.**

2. **The legislative and regulatory framework shall provide for:**

   (i) **the establishment of applicable national safety requirements and regulations**

   (ii) **a system of licensing with regard to nuclear installations and the prohibition of the operation of a nuclear installation without a license**

   (iii) **a system of regulatory inspection and assessment of nuclear installations to ascertain compliance with applicable regulations and the terms of licenses**

   (iv) **the enforcement of applicable regulations and of the terms of licenses, including suspension, modification, or revocation**

This section explains the legislative and regulatory framework governing the U.S. nuclear industry. It discusses the provisions of that framework for establishing national safety requirements and regulations and systems for licensing, inspection, and enforcement.

7.1 Legislative and Regulatory Framework

The Atomic Energy Act of 1954, passed by Congress and signed by the President, established the Atomic Energy Commission and the legal framework for all subsequent regulation of nuclear installations. However, as is generally the case with most laws, this act provided general principles and concepts and left the regulatory body (i.e., the NRC) to address the details through specific regulations. The Energy Reorganization Act of 1974, likewise passed by Congress and signed by the President, abolished the Atomic Energy Commission and created the NRC to regulate commercial nuclear activities and the U.S. Energy Research and Development Administration (ERDA) to continue government-sponsored nuclear activities. ERDA was subsequently incorporated into the U.S. Department of Energy (DOE). The Administrative Procedure Act provides the general rules and procedures through which the Atomic Energy Act is implemented.

The United States has also ratified various international conventions that impact nuclear safety:

- The Nuclear Non-Proliferation Treaty, ratified in 1970, governs the NRC's export licensing activities.
- The U.S.-IAEA Safeguards Agreement, ratified in 1980, requires eligible facilities in the United States to report material accounting data on declared nuclear material. The Agreement further requires eligible facilities to submit to IAEA inspections. The Additional Protocol to the US-IAEA Safeguards Agreement, ratified in 2004, strengthened IAEA reporting and access rights for eligible facilities.
- The Convention on the Physical Protection of Nuclear Material, ratified in 1982, requires NRC licensees to take steps to protect nuclear material during international transport.

55

- The Convention on Early Notification of a Nuclear Accident, ratified in 1988, requires the NRC to help the U.S. Department of State report significant accidents to IAEA and any State affected by a transboundary radioactive release.
- The Convention on Assistance in the Case of a Nuclear Accident or Radiological Emergency, ratified in 1988, requires the NRC to help the U.S. Department of State respond to requests for assistance in the event of a foreign nuclear accident or emergency.
- The Joint Convention on the Safety of Spent Fuel Management and on the Safety of Radioactive Waste Management ("Joint Convention"), ratified in 2003, requires the United States to take steps to ensure that individuals and the environment are protected against radiological hazards at all stages of radioactive waste and spent fuel management. The Joint Convention further calls for periodic review meetings of all the Contracting Parties. Before the review meeting, each Contracting Party must submit a national report that addresses measures taken to implement the obligations under the Joint Convention.
- The Convention on Supplementary Compensation for Nuclear Damage, ratified in 2008, requires the United States to ensure that adequate compensation exists in the event that "nuclear damage" results from a nuclear incident.
- The Convention on Nuclear Safety, ratified in 1999, requires the United States to submit periodic National Reports that detail the United States' commitment to nuclear safety.

7.2 Provisions of the Legislative and Regulatory Framework

7.2.1 National Safety Requirements and Regulations

In addition to the Atomic Energy Act, several statutes (listed in previous U.S. National Reports) have substantial bearing on the practices and procedures of the Commission. Furthermore, various U.S. Presidents have issued Executive orders and directives that impact nuclear safety. For example, President Reagan issued Executive Order 12656 "Assignment of Emergency Preparedness Responsibilities," on November 18, 1988. This Executive order assigned certain emergency preparedness responsibilities to the NRC in case of a national emergency. Likewise, in the wake of the Three Mile Island accident, President Carter directed the Federal Emergency Management Agency (FEMA) to lead up all off-site emergency activities and review emergency plans in States with operating reactors. As a final example, the NRC has voluntarily complied with President Clinton's Executive Order 12898, "Federal Actions To Address Environmental Justice in Minority Populations and Low-Income Populations," dated February 11, 1994, which required agencies to consider whether its programs or policies have a disproportionately adverse health or environmental effect on minority populations.

The NRC has implemented these statutes and Executive orders through regulation. Specifically, Title 10 of the *Code of Federal Regulations*, "Energy," governs the licensing of nuclear installations. The NRC established these regulations through "notice-and-comment" rulemaking procedures under the Administrative Procedure Act. In short, these procedures include: (1) establishing a technical or legal basis, or both, for the proposed rule, (2) inviting interested parties to comment on the proposed rule, and (3) considering comments and issuing a final rule. Once these final rules are in place, they are binding on operators of nuclear installations and can be revised only through a new notice-and-comment rulemaking. This ensures that interested parties remain both informed of, and involved with, any changes to the NRC's regulatory scheme.

7.2.2 Licensing of Nuclear Installations

The NRC must license all commercial nuclear installations in the United States.   (Some Government facilities that are operated by or for DOE are not subject to NRC licensing under the Atomic Energy Act and the Energy Reorganization Act except where specifically provided by law).   The Atomic Energy Act, Chapter 10, Section 101, prohibits operation of a nuclear installation without a valid license.   Sections 101 and 103 further provide that only the NRC is authorized to issue a license for nuclear reactor facilities.   Section 103 also states that such licenses are subject to conditions that the NRC may establish by rule or regulation to carry out the purposes and provisions of the Atomic Energy Act.

The Atomic Energy Act, Section 189a, provides interested parties with hearing rights in proceedings for the granting, suspending, revoking, or amending of a license or construction permit.   Hearings, which are used in licensing proceedings for production and utilization facilities (e.g., nuclear power plants), are held under procedural rules stated in 10 CFR Part 2, "Rules of Practice for Domestic Licensing Proceedings and Issuance of Orders," and, in particular, Subpart C, "Rules of General Applicability."   The NRC staff participates as a party in most formal hearings and may also participate as a party in less formal hearings.   Hearings are usually held before a three-member Atomic Safety and Licensing Board, which is generally composed of one lawyer and two technical members.

Article 18 of this report describes the licensing process in greater detail.   Two alternative approaches to licensing exist.   The traditional approach, under 10 CFR Part 50, requires two steps.   In the first step, the NRC reviews a preliminary application and decides whether to grant a construction permit.   In the second step, the agency reviews the final application and decides whether to grant an operating license.   The NRC licensed all current operating plants in the United States according to this process.

In 1989, the Commission established an alternative licensing system, published in 10 CFR Part 52, which provides for certified standard designs and combined licenses that resolve design issues before construction, and early site permits that resolve most siting issues years before construction.   The basic concept underlying 10 CFR Part 52 is that the NRC can approve nuclear reactor designs through generic rulemaking.   Once the designs are approved, an applicant can reference them in applications for permission to build and operate nuclear power plants without needing to relitigate, in individual hearings, the issues resolved in the rulemaking.   Moreover, the NRC will determine and approve before construction the criteria for evaluating whether the plant had been built as specified.   Thus, the plant could begin operation without a second hearing, provided that it satisfied the acceptance criteria.   To the extent possible, issues would be litigated before construction, not once construction is nearly complete, when the consequences of delay are much greater.   In adopting 10 CFR Part 52, the Commission used the latitude allowed by law to streamline licensing.

Recently, the NRC amended 10 CFR Part 52 to improve the effectiveness of its processes for licensing future nuclear power plants.   The amendments clarify the overall regulatory relationship between 10 CFR Part 50 and 10 CFR Part 52, reorganize 10 CFR Part 52, and reconcile differences in wording in other parts of the regulations to provide consistent terminology throughout all of the regulations affecting 10 CFR Part 52.   The amendments also added new sections on written communications, employee protection, completeness and accuracy of information, exemptions, combining licenses, and jurisdictional limits.

Once licensed, a nuclear power plant can renew its operating license for up to an additional 20 years. The NRC provides the licensing system for license renewal under 10 CFR Part 54, "Requirements for Renewal of Operating Licenses for Nuclear Power Plants," and interested parties have hearing rights under 10 CFR Part 2 in renewal proceedings.

7.2.3 Inspection and Assessment

Under the Atomic Energy Act, the NRC has the authority to inspect nuclear power plants in its role of protecting public health and safety and the common defense and security. The NRC staff inspects power reactors under construction, in test conditions, and in operation to ascertain compliance with regulations and license conditions. Through its inspection program, the NRC assesses whether activities are properly conducted and equipment is properly maintained to ensure safe operations. The agency integrates inspection results into its overall evaluation of licensee performance, as discussed in Article 6 of this report. If a safety problem exists, or there is a failure to comply with requirements, the licensee must take prompt corrective action. If necessary, the NRC may take enforcement action.

A primary feature of the NRC's inspection program is the assignment of resident inspectors to nuclear power plants. At least two inspectors are assigned to each nuclear power site, and these inspectors continuously monitor licensee activities in accordance with the NRC's baseline inspection program. To supplement these continuous inspections, regional inspection specialists conduct periodic inspections of each plant in his or her region. If needed, regional inspectors perform special investigations of plants that exceed established thresholds during routine inspections and thus require heightened scrutiny. All inspection findings are recorded, and the NRC typically issues inspection reports for a specific power plant each quarter. Additionally, senior agency managers review plants that have performance issues and report these results to the Commission during the annual Agency Action Review Meeting. This meeting provides another opportunity to discuss significant events, licensee performance issues, trends, and the actions to mitigate recurrences.

7.2.4 Enforcement

The NRC draws its jurisdiction for enforcement from the Atomic Energy Act and the Energy Reorganization Act.

The Atomic Energy Act, Section 161, authorizes the NRC to conduct inspections and investigations and to issue orders as may be necessary or desirable to promote the common defense and security, protect health, or minimize danger to life or property. Section 186 authorizes the NRC to revoke licenses under certain circumstances (e.g., for material false statements, for a change in conditions that would have warranted NRC refusal to grant a license on an original application, for a licensee's failure to build or operate a facility in accordance with the terms of the permit or license, and for violation of an NRC regulation). Section 234 authorizes the NRC to impose monetary civil penalties not to exceed $100,000 per violation per day; however, that amount is adjusted every 4 years by the Federal Civil Penalties Inflation Adjustment Act of 1990, as amended by the Debt Collection Improvement Act of 1996, and is currently $140,000. In addition to the provisions mentioned in Section 234, Sections 84 and 147 authorize the imposition of civil penalties for violations of regulations implementing those provisions. Section 232 authorizes the NRC to seek injunctive or other equitable relief for violation of regulatory requirements.

The Atomic Energy Act, Chapter 18, provides for varying levels of criminal penalties (i.e., monetary fines and imprisonment) for willful violations of the act or the regulations or orders issued under Sections 65, 161b, 161i, or 161o of the act. Section 223 allows the NRC to impose criminal penalties on certain individuals who are employed by firms constructing or supplying basic components of any utilization facility if the individual knowingly and willfully violates NRC requirements in a manner that could significantly impair a basic component. Section 235 allows the NRC to impose criminal penalties on persons who interfere with nuclear inspectors. Section 236 allows the NRC to impose criminal penalties on persons who attempt to or cause sabotage at a nuclear facility or to nuclear fuel. The agency refers alleged or suspected instances of criminal violations of the Atomic Energy Act to the U.S. Department of Justice for appropriate action.

The Energy Reorganization Act, Section 206, authorizes the NRC to impose civil penalties on licensees for knowing and conscious failures to provide the agency with certain safety information.

Subpart B, "Procedure for Imposing Requirements by Order, or for Modification, Suspension, or Revocation of a License, or for Imposing Civil Penalties," of 10 CFR Part 2 specifies the procedures that the NRC uses in exercising its enforcement authority. The scope of Subpart B includes the following procedures:

- 10 CFR 2.201, "Notice of Violation," outlines the procedure for issuing notices of violations.

- 10 CFR 2.202, "Orders," explains the procedure for issuing orders. In accordance with this section, the NRC may decide to issue an order to institute a proceeding to modify, suspend, or revoke a license or to take other action against a licensee or other person subject to the NRC's jurisdiction. The licensee or any other person adversely affected by the order may request a hearing. The NRC is authorized to make orders immediately effective if required to protect public health, safety, or interest, or if the violation is willful.

- 10 CFR 2.204, "Demand for Information," specifies the procedure for issuing a demand for information to a licensee or other person subject to the Commission's jurisdiction to determine whether an order should be issued or other enforcement action should be taken. The demand does not provide hearing rights because the agency is only seeking information. A licensee must answer a demand for information. An unlicensed person may answer a demand either by providing the requested information or by explaining why the demand should not have been issued.

- 10 CFR 2.205, "Civil Penalties," describes the procedure for assessing civil penalties. The NRC initiates the civil penalty process by issuing a notice of violation and proposed imposition of a civil penalty. The agency provides the person charged with an opportunity to contest in writing the proposed imposition of a civil penalty. After evaluating the response, the NRC may mitigate, remit, or impose the civil penalty. If the agency imposes a civil penalty, it provides an opportunity for a hearing. If a civil penalty is not paid following a hearing, or if a hearing is not requested, the agency may refer the matter to the U.S. Department of Justice to institute a civil action in Federal district court to collect the penalty.

The NRC has had positive experience with legal actions and enforcement measures.   As noted in Section 9.3 of this report, the NRC has recently undertaken many successful enforcement actions against licensees.   These actions are rarely challenged before an Atomic Safety and Licensing Board, and the appellate courts rarely overturn the NRC's decision.

ARTICLE 8. REGULATORY BODY

1. **Each Contracting Party shall establish or designate a regulatory body entrusted with the implementation of the legislative and regulatory framework referred to in Article 7, and provided with adequate authority, competence, and financial and human resources to fulfill its assigned responsibilities.**

2. **Each Contracting Party shall take the appropriate steps to ensure an effective separation between the functions of the regulatory body and those of any other body or organization concerned with the promotion or utilization of nuclear energy.**

This section explains the establishment of the U.S. regulatory body (i.e., the NRC). It also explains how the functions of the NRC are separate from those of bodies responsible for promoting research, development and advancement of nuclear energy (e.g., DOE).

8.1 The Regulatory Body

This section explains the NRC's mandate, authority and responsibilities, structure, international responsibilities and activities, financial and human resources, position in the governmental structure, and report of the Integrated Regulatory Review Service (IRRS) self-assessment team.

8.1.1 Mandate

As discussed in Article 7, Congress created the NRC as an independent regulatory agency in January 1975, with the passage of the Energy Reorganization Act. In giving the NRC an exclusively regulatory mandate, the statute reflected (in part) a congressional judgment that the expanding commercial nuclear power industry (which was expected to continue to grow) warranted the full-time attention of an exclusively regulatory agency. In creating the NRC, Congress also addressed a developing public concern that regulatory responsibilities were overshadowed by the promotion of nuclear power at the Atomic Energy Commission.

8.1.2 Authority and Responsibilities

8.1.2.1 Scope of Authority

The NRC's mission is to ensure that the civilian uses of nuclear energy and materials in the United States are conducted with proper regard for public health and safety, national security, and environmental concerns. The Atomic Energy Act provides the charter for these regulatory responsibilities through which the U.S. Congress created a national policy of developing the peaceful uses of atomic energy. The U.S. Congress has amended the statute over the years to address developing technology and changing regulatory needs. For example, the National Environmental Policy Act of 1969, as amended, imposed broad new responsibilities on Federal agencies. Other more specialized statutes prescribe the NRC's duties with regard to high-level radioactive waste, low-level radioactive waste, mill tailings, environmental reviews, nonproliferation, antiterrorism, and import and export of nuclear materials and equipment.

The NRC's licensing authority extends to other Government organizations (such as the Tennessee Valley Authority (TVA), which operates nuclear power plants) and to the military's use of radiopharmaceuticals in its hospitals, but not to the military's or the DOE's nuclear weapons programs and facilities, nor to the DOE's test and research reactors.   The NRC's responsibilities include ensuring both safety and the security of commercial nuclear facilities and materials against radiological sabotage and thefts.

In addition, the NRC is authorized to relinquish its authority over certain matters to States (i.e., of the United States) that enter into agreements with the NRC.   Such States are known as Agreement States.

Section 8.2 of this report provides specific information about the scope of the agency's limited authority over DOE nuclear installations.

8.1.2.2   The NRC as an Independent Regulatory Agency

The Commission's status as an independent regulatory agency within the executive branch of the Federal Government means that the President cannot ordinarily direct its regulatory decisions.   There are two statutory sources of the Commission's independence from presidential direction.   First, the President can remove an NRC Commissioner only for cause – namely, "inefficiency, neglect of duty, or malfeasance in office."   The President can, however, designate one member of the Commission to serve as Chairman.   Second, the Commission has the statutory right to defend itself whenever its safety findings are challenged in U.S. appellate courts.

Congress cannot override the Commission's decisions, except by duly enacted legislation.   The courts are likewise limited in reviewing the NRC's factual safety findings.   Although a Federal appellate court can overturn a Commission decision for violations of law, safety findings will generally be overturned only if they are arbitrary.   This provides the Commission with some degree of independence from a court's second-guessing the NRC's technical factual findings.

The independence of the NRC's decisionmaking process implies a responsibility on the part of the Commissioners and their personal staffs to keep the process free from improper outside influence.   This is especially important in the case of adjudications.   When the Commissioners take part in adjudications, they ordinarily act in the role of appellate judges (reviewing the decisions of lower judges) and, in general, are bound by the same kinds of strictures that apply to judges in Federal courts.

8.1.3   Structure of the Regulatory Body

This section explains the structure of the NRC.   It covers the Commission, component offices and their responsibilities, and advisory committees and their functions.   It also explains recent changes in NRC organization.

8.1.3.1   The Commission

The NRC is headed by a five-member Commission.   The President designates one member to serve as Chairman and official spokesperson.   The Commission as a whole formulates policies and regulations governing nuclear reactor and materials safety, issues orders to licensees, and adjudicates legal matters brought before it.   The Executive Director for Operations carries out the policies and decisions of the Commission and directs the activities of the program offices.

8.1.3.2 Component Offices of the Commission

The following offices report directly to the Chairman or the Commission:

- **Office of the Executive Director for Operations**. The Executive Director for Operations is the chief operational and administrative officer of the Commission and is authorized and directed to discharge such licensing, regulatory, and administrative functions and to take such actions as necessary for day-to-day agency operations. The Executive Director for Operations supervises and coordinates the policy development and operational activities of the NRC program and regional offices and implements Commission policy directives pertaining to these offices.

- **Office of the Chief Financial Officer**. The Office of the Chief Financial Officer is responsible for the NRC's planning, budgeting, and performance management process and for all NRC financial management activities.

- **Office of the General Counsel**. The Office of the General Counsel directs matters of law and legal policy, providing opinions, advice, and assistance to the agency on all of its activities.

- **Office of the Inspector General**. The Inspector General provides leadership and policy direction in conducting audits and investigations to promote economy, efficiency, and effectiveness within the NRC and to prevent and detect fraud, waste, abuse, and mismanagement in agency programs and operations.

- **Office of International Programs**. The Office of International Programs coordinates the NRC's international activities and provides assistance and recommendations to the Chairman, the Commission, and the NRC staff. It plans, develops, and implements programs to carry out policies in the international arena, including export and import licensing responsibilities. It establishes and maintains working relationships with individual countries and international nuclear organizations, as well as other involved U.S. Government agencies.

- **Office of Public Affairs**. The Office of Public Affairs directs the agency's public affairs program, advising agency officials and developing key strategies that help increase public confidence in NRC policies and activities.

- **Office of Congressional Affairs**. The Office of Congressional Affairs is the primary point of contact for all communications between the NRC and Congress. This office provides advice and assistance to the Chairman, the Commission, and NRC staff on congressional matters; monitors legislative proposals, bills, and hearings; informs the NRC of the views of Congress on NRC policies, plans, and activities; provides timely responses to congressional requests for information; and provides the information necessary to keep appropriate Members of Congress and congressional staff fully and currently informed of NRC actions.

- **The Office of Commission Appellate Adjudication**. The Office of Commission Appellate Adjudication provides the Commission with an analysis of any adjudicatory matter requiring a Commission decision and drafts necessary decisions pursuant to the Commission's guidance after a presentation of options.

- <u>Office of the Secretary of the Commission</u>. The Office of the Secretary of the Commission provides executive management services to support the Commission and to carry out Commission decisions. It assists with the planning, scheduling, and conduct of Commission business; maintains historical paper files of official Commission records; administers the NRC Historical Program; and maintains the Commission's official adjudicatory and rulemaking dockets.

8.1.3.3 Offices of the Executive Director for Operations

The offices reporting to the Executive Director for Operations ensure that the commercial use of nuclear materials in the United States is safely conducted. Since issuance of the previous U.S. National Report, the NRC established a new office, the Computer Security Office, which was operational in November 2007. The following briefly describes this and other NRC offices:

- <u>Office of Nuclear Reactor Regulation</u>. The Office of Nuclear Reactor Regulation is responsible for accomplishing key components of the NRC's nuclear reactor safety mission to protect public health and safety and the environment. To do so, the office conducts a broad range of regulatory activities in the four primary program areas of rulemaking, licensing, oversight, and incident response for commercial nuclear power reactors and test and research reactors.

- <u>Office of New Reactors</u>. The Office of New Reactors is responsible for accomplishing key components of the NRC's nuclear reactor safety mission for new reactor facilities licensed in accordance with 10 CFR Part 52. As such, the office conducts regulatory activities in the primary program areas of siting, licensing, and oversight for new commercial nuclear power reactors.

- <u>Office of Nuclear Material Safety and Safeguards</u>. The Office of Nuclear Material Safety and Safeguards is responsible for regulating activities that provide for the safe and secure production of nuclear fuel used in commercial nuclear reactors; the safe storage, transportation, and disposal of high-level radioactive waste and spent nuclear fuel; and the transportation of radioactive materials regulated under the Atomic Energy Act.

- <u>Office of Nuclear Security and Incident Response</u>. The Office of Nuclear Security and Incident Response develops overall agency policy and provides management direction for evaluating and assessing technical issues involving security and emergency preparedness at nuclear facilities.

- <u>Office of Nuclear Regulatory Research</u>. The Office of Nuclear Regulatory Research plans, recommends, and conducts research programs and technical safety reviews that support the resolution of ongoing and future safety issues identified as regulatory needs by offices with regulatory functions or through its own long-term research review program.

- <u>Office of Enforcement</u>. The Office of Enforcement oversees, manages, and directs the development and implementation of policies and programs for enforcing NRC requirements. It oversees the agency's allegations management program and the allegations review process. The office is responsible for external safety culture policy matters, the agency's Alternative Dispute Resolution Program, the agency's internal Differing Professional Opinions Program, and its internal non-concurrence process.

- Office of Investigations. The Office of Investigations develops policy, procedures, and quality control standards for investigations of licensees and applicants, as well as their contractors or vendors. This office conducts investigations of allegations of wrongdoing by non-NRC employees and contractors. The Office of Investigations is independent and may self-initiate investigations when a person or entity under its jurisdiction is suspected to have committed a matter of wrongdoing. This office plans, conducts, and makes referrals of substantiated criminal cases to the U.S. Department of Justice. This office conducts liaison with Federal, State, and local law enforcement and provides investigative assistance to NRC staff on regulatory matters. Additionally, it keeps the Commission and NRC offices apprised of regulatory matters under investigation as they affect public health and safety, the common defense and security, and the environment.

- Office of Federal and State Materials and Environmental Management Programs. The Office of Federal and State Materials and Environmental Management Programs is responsible for the safe and secure use of source, byproduct, and special nuclear materials in industrial, medical, academic, and commercial activities, and at decommissioning, uranium recovery, and low-level waste sites. It ensures effective communications and working relationships between the NRC and other governmental entities and administers the Agreement State Program (through which States have signed formal agreements with the NRC to assume regulatory responsibility over certain byproduct, source, and small quantities of special nuclear materials). It also develops and implements rules and guidance for these activities.

- Office of Information Services. The Office of Information Services plans, directs, and oversees the delivery of centralized information technology infrastructure, applications, and information management services, in addition to the development and implementation of information technology and management plans, architecture, and policies to support the mission, goals, and priorities of the agency.

- Regional Offices. The four regional offices conduct inspections, and execute established policies related to licensing and construction, allegation, enforcement, emergency response, and government liaison programs in the United States licensed nuclear facilities. The regional offices also manage decommissioning activities.

Supporting the Executive Director for Operations are the Offices of Administration, Human Resources, Small Business and Civil Rights, and Computer Security:

- Office of Administration. The Office of Administration provides centralized services in the areas of contracts, facilities and security, property management, and administration, including support for rulemaking and agency directives, transportation, parking, translations, audiovisual needs, food services, mail distribution, labor services, furniture and supplies, and other areas.

- Office of Human Resources. The Office of Human Resources provides overall leadership and management of the agency's human capital planning and training and development programs.   Accordingly, this office is responsible for implementing human resource policy and operations agency-wide.   This includes overseeing the development and implementation of human resources management and information systems for staffing, strategic workforce planning, and other corporate activities to support a skilled and dynamic workforce.   The office's training and development programs are designed to establish, maintain, and enhance the skills employees need today and to meet the agency's future skill needs.

- <u>Office of Small Business and Civil Rights</u>.   The Office of Small Business and Civil Rights is responsible for facilitating equal employment opportunity for all NRC employees, applicants for employment, and business partners through an on-going affirmative employment and diversity management process, implementation of civil rights statues, execution of outreach and compliance coordination mandates, and employment of maximum small business participation in acquisitions.

- <u>Computer Security Office</u>.   The Computer Security Office plans, directs, and oversees the implementation of a comprehensive, coordinated, integrated and cost-effective NRC information technology security program, consistent with applicable laws, regulations, and Commission, Executive Director for Operations, and Chief Information Officer direction, management initiatives, and policies.

8.1.3.4   Advisory Committees

The three principal advisory committees for NRC programs are the Advisory Committee on Reactor Safeguards, the Advisory Committee on Medical Uses of Isotopes and the Committee to Review Generic Requirements.   In addition, the NRC has established an ad hoc Licensing Support Network Advisory Panel.   Most relevant to this report are the Advisory Committee on Reactor Safeguards and the Committee to Review Generic Requirements.   The Advisory Committee on Reactor Safeguards reviews and reports on safety studies and reactor facility license and license renewal applications, advises the Commission on the hazards of proposed and existing reactor facilities and the adequacy of proposed reactor safety standards, advises the Commission on issues associated with nuclear materials and waste management, initiates reviews of specific generic matters or nuclear facility safety-related items, and reviews the NRC's research activities.   The Committee to Review Generic Requirements ensures that proposed generic backfits to be imposed on NRC-licensed power reactors and selected nuclear materials licensees are appropriately justified, based on the backfit provisions of applicable NRC regulations and the Commission's backfit policy.

8.1.3.5   Atomic Safety and Licensing Board Panel

In addition to the advisory committees, the NRC has an Atomic Safety and Licensing Board Panel.   This panel conducts hearings for the Commission and performs such other regulatory functions as the Commission authorizes.   The Chief Administrative Judge develops and applies procedures governing the activities of boards, administrative judges, and administrative law judges.   This person also makes appropriate recommendations to the Commission concerning the rules governing the conduct of hearings.

8.1.4 International Responsibilities and Activities

The NRC conducts international activities related to statutory mandates, international treaties and conventions, international organizations, bilateral relations, and research.

U.S. law or international treaties and conventions mandate several NRC international activities; other activities are discretionary. In particular, the NRC is statutorily mandated to serve as the U.S. licensing authority for exports and imports of nuclear materials and equipment.

The NRC supports U.S. foreign policy in the safe and secure use of nuclear materials and in guarding against the spread of nuclear weapons. The agency actively participates in developing and implementing a variety of legally binding treaties and conventions that create an international framework for the peaceful uses of nuclear energy. The NRC provides technical and legal advice and assistance to international organizations and foreign countries as they work to develop effective regulatory organizations and rigorous safety standards. Some activities are carried out within the programs of IAEA, the Nuclear Energy Agency (NEA) of the Organisation for Economic Co-operation and Development, or other international organizations. The NRC conducts other activities directly with counterpart agencies in other countries under cooperation agreements.

International Treaties. Treaties that legally bind the NRC and the U.S. Government's peaceful uses of nuclear energy and nuclear applications include the 1978 Nuclear Non-Proliferation Treaty, the 1980 Convention on Physical Protection of Nuclear Material, the 1994 Convention on Nuclear Safety, the 1986 Convention on Early Notification of a Nuclear Accident, the 1986 Convention on Assistance in Case of a Nuclear Accident or Radiological Emergency, and the 1997 Joint Convention on the Safety of Spent Fuel Management and on the Safety of Radioactive Waste Management. NRC staff members regularly participate in international meetings related to these conventions and have held a variety of convention leadership positions. In its bilateral work with regulatory counterparts worldwide, the NRC seeks to exchange experience and good practices in order to further the goals of these international instruments.

In addition to these legally-binding obligations, the United States has agreed to comply with certain activities to enhance the safe and secure uses of nuclear applications. For example, the U.S. has made a political commitment to implement the IAEA Code of Conduct on the Safety and Security of Radioactive Sources. This commitment has been codified in U.S. statute as part of the Energy Policy Act of 2005 and is reflected in the NRC's export and import regulations.

Export-Import. The NRC's key international responsibility is licensing the export and import of nuclear materials and equipment for civilian use, such as low-enriched uranium fuel for nuclear power plants, high-enriched uranium for research and test reactors, nuclear reactors themselves, certain nuclear reactor components (such as pumps and valves), and radioisotopes used in industrial, medical, agricultural, and scientific fields. The NRC ensures that such exports and imports are consistent with the goals of the safe and peaceful use of these materials and equipment, limiting the proliferation of nuclear weapons, and promoting the Nation's common defense and security. The Atomic Energy Act, the Nuclear Non-Proliferation Act of 1978, and 10 CFR Part 110, "Export and Import of Nuclear Equipment and Material," detail the standards and procedures for issuing export and import licenses. The NRC also coordinates closely with other U.S. Government agencies, including the U.S. Department of State, U.S. Department of Commerce, and DOE, on export- or import-related matters that fall within these agencies' jurisdictions.

International Organizations and Associations.   In consultation with the executive branch agencies, the NRC actively participates in the full scope of programs of the two major international nuclear organizations, IAEA and NEA.   For example, since 1996, the United States has or is planning to participate in more than 30 Operational Safety Assessment Review Team (OSART) missions.   Some experts on these teams come from the NRC, while others come from industry.   The NRC coordinates closely with the Institute of Nuclear Power Operations (INPO) in this process.   The NRC is currently working with IAEA and industry in planning an OSART mission to Seabrook Unit 1 in 2011 and intends to continue to plan for an OSART mission in the United States every 3 years.   Since 1999, the NRC has participated in more than 20 Integrated Regulatory Review Teams or IRRS missions, sending high-level technical experts on approximately four missions per year.   In October 2010, the United States will host an IRRS mission, focused on the U.S. operating reactor program.

The NRC holds leadership roles in the four IAEA Safety Standards Committees and the Commission on Safety Standards.   These activities, together with regular NRC staff participation in IAEA meetings to draft and revise safety and security guidance in coordination with other U.S. Government agencies, enable the NRC to use its broad regulatory experience to contribute to the safe and secure use of nuclear and radioactive materials in IAEA Member States.

The NRC also participates in the NEA Steering Committee, and holds leadership positions on NEA's Committee on the Safety of Nuclear Installations, the Committee on Nuclear Regulatory Activities, the Committee on Radiation Protection and Public Health, and the Radioactive Waste Management Committee.   The NRC also holds leadership roles in, and is otherwise represented on, many of the NEA committee-chartered working groups.   These activities provide diverse forums for nuclear regulators and research organizations to share information and work together to leverage resources for mutual benefit.

The NRC continues to participate in the Multinational Design Evaluation Program, with the goal of leveraging the experience of international counterparts in the review of new reactor designs. Through this program, the NRC is (1) sharing information with other regulatory authorities in the reviews of the Westinghouse's Advanced Passive (AP) 1000 and AREVA Nuclear Power's U.S. Evolutionary Power Reactor (US EPR) designs, (2) cooperating in vendor inspections, and (3) pursuing possible convergence of regulations, codes, and standards associated with the design reviews of new reactors.

The NRC has been working closely with IAEA in support of countries seeking to develop new nuclear power programs or expand small or dormant programs.   The NRC staff has been active in guidance document development in this area and has participated in numerous workshops and training activities to provide so-called "new entrant" countries with information and experience on building a robust, independent, regulatory infrastructure.   In 2010, the NRC provided a cost-free expert to assist IAEA in its activities in this area.   The NRC also funded a comparable position at NEA to assist in identifying how NEA's work, within its focused membership, may benefit countries with more established technical and regulatory programs. The NRC also works closely with its international regulatory counterparts, through such mechanisms as the IAEA-sponsored Regulatory Cooperation Forum, to coordinate efforts for supporting regulatory development in the "new entrant" countries.

In addition to staff participation in more than 100 IAEA and NEA meetings each year, the NRC Chairman routinely participates in the IAEA General Conference and biannual meetings of the International Nuclear Regulators Association.   Members of the Commission also travel to international conferences around the world to deliver keynote remarks, participate in panel discussions, and otherwise share insight on a variety of topics with diverse international audiences.

Bilateral Relations.   The NRC works closely with nuclear safety agencies in more than 40 countries.   The NRC and its foreign counterparts routinely exchange operational safety data and other regulatory information.   The NRC provides safety, security, emergency preparedness and safeguards advice, training, and other assistance to countries that seek U.S. help to improve their regulatory programs.

The NRC's information exchange arrangements serve as communication channels with foreign regulatory authorities, establishing a framework for the agency to gain access to non-U.S. safety information that can (1) alert the U.S. Government and industry to potential safety problems, (2) help identify possible accident precursors, and (3) provide accident and incident analyses, including lessons learned, that could be directly applicable to the safety of U.S. nuclear power plants and other facilities.   The arrangements also serve as vehicles for the assistance the NRC provides to countries to establish and improve their regulatory capabilities and infrastructure. Thus, the arrangements facilitate the NRC's strategic goal to support U.S. interests in the safe and secure use of nuclear materials and in nuclear nonproliferation.   The NRC currently has 38 active bilateral arrangements with its foreign regulatory counterparts.   These arrangements allow the staff to conduct regular bilateral exchanges on a variety of levels.   The NRC also has bilateral interactions with countries with which there is no active arrangement, although the absence of an arrangement limits the type of information that can be exchanged.   The NRC Chairman typically meets with at least 20 foreign counterparts at IAEA's annual General Conference.   In addition, members of the Commission travel abroad to hold bilateral meetings with their regulatory counterparts, tour nuclear power plants and other facilities, and exchange information and good practices.   Often, these visits result in increased communication between the NRC and its counterparts, providing opportunities for enhanced information exchange based on first-hand knowledge of various programs.

International Assistance Programs.   In the early 1990s, the NRC began offering assistance to nuclear regulatory programs in several former Soviet states.   The agency initially focused its efforts on those countries in which Soviet-designed reactors were operated.   Following the September 11, 2001, terrorist attacks, the NRC expanded its assistance efforts to specifically include assisting countries in their efforts to improve regulatory oversight of radioactive sources. In addition, the NRC is assisting the Government of Iraq in its efforts to develop a sound regulatory structure, including the provision of assistance in developing the law and regulations that will be the legal framework for the project of decommissioning former Iraqi facilities that used radioactive materials.   The NRC is also providing bilateral assistance to countries seeking to establish nuclear power programs, in close consultation with IAEA.   IAEA and the U.S. Government are both actively promoting regional cooperation, and have been engaged in workshops and training activities to further that goal.

Research Programs.   The NRC conducts confirmatory regulatory research through the implementation of more than 100 bilateral and multilateral agreements in partnership with nuclear safety agencies and institutes in more than 30 countries.   This research supports regulatory decisions on emerging technologies, aging equipment and facilities, and various other safety issues.   The NRC and other nuclear regulatory and safety organizations carry out cooperative research projects to achieve mutual research needs with greater efficiency.

8.1.5 Financial and Human Resources

8.1.5.1 Financial Resources

As of September 30, 2009, the NRC's financial condition was sound in that the agency had sufficient funds to meet program needs and adequate control of these funds in place to ensure that obligations did not exceed budget authority.   The sum of all funds available to obligate for FY 2009 was $1,165.2 million, which is a $136.4 million increase over the FY 2008 amount of $1,028.8 million

The NRC FY 2010 budget was financed with $912.2 million from user fees, $125.7 million from the General Fund, and $29 million from the Nuclear Waste Fund.   The NRC FY 2011 proposed budget will be financed with $915.3 million from user fees, $128.3 million from the General Fund, and $10.0 million from the Nuclear Waste Fund.

8.1.5.2 Human Resources

The NRC worked aggressively to hire the highly skilled staff needed to regulate the existing fleet of operating nuclear reactors and to meet the demands of new reactor and materials license application reviews.   The NRC is now hiring at a slower pace.   For example, in 2008, the agency hired more than 500 new employees, while in 2009 it brought in 287 new employees, and approximately half of these were hired to replace staff members who left.   The NRC is now working diligently to meet the challenge of training and integrating a new and increasingly younger workforce, providing them with the necessary infrastructure to successfully carry out the organization's mission.

Responses to employee viewpoint surveys show that the agency is on the right path.   In 2007 and 2009, the NRC was ranked as the best place to work in the Federal Government.   The results of the 2009 survey reflect that employees feel strongly engaged, understand how their work contributes to the agency's mission, and view their work as meaningful and important. Survey results also indicated that employees agree that they have the training, development, information, and skills needed to perform their work.   The safety culture survey conducted by the Office of the Inspector General similarly reflected positive employee perceptions, even when compared to organizations viewed as the best in class.   The NRC continues to use such surveys to choose areas for further focus and improvement in its management of human resources.   For example, the NRC implemented initiatives to (1) ensure that all employees understand the relevance to their work of an open, collaborative working environment (OCWE) and strong safety culture, (2) further communicate information about benefits to all staff members, (3) enhance work-life flexibilities such as telecommuting and flexible work schedules, and (4) continuously improve performance management and communications.

Recruitment and Hiring Process.   To meet current hiring demands and to increase efficiency in hiring, the NRC identified the need to focus its recruitment activities and streamline the hiring process.   As a long-standing practice, the NRC actively recruits for its Nuclear Safety Professional Development Program at targeted universities with a history of graduating technically strong, diverse candidates.   In addition, the NRC has maintained its recruitment activities at professional society conferences and career fairs.   The agency advertises in trade journals and on Web sites to attract professionals in specialized technical disciplines and in local newspapers around the country in areas where technical engineers and scientists may be interested in re-locating because of job cutbacks.

The agency continued to make prudent, targeted use of recruitment, relocation, and retention incentives and pension offset waiver (rehiring annuitants without reduction of salary or pension) in order to hire and retain employees of the quality necessary to carry out the agency's mission. Such incentives are particularly useful for unusual occupations or highly specialized disciplines for which candidates may be scarce. The NRC offers non-supervisory employees referral awards when they are instrumental in helping the agency fill positions. The NRC also continues to strengthen its programs for developing and hiring students in critical specialties through programs such as partnerships with colleges and universities; university grants, scholarships, and fellowships; cooperative education programs; and payment of transportation and lodging expenses for student employees.

Training and Knowledge Management and Transfer. Nearly half of NRC staff members have been with the agency for less than 6 years. Rapidly training and integrating this large number of new employees into the agency is a significant challenge, but it is essential for the NRC's and the employees' future success and productivity. To address this challenge, the NRC is expanding the use of existing learning tools, including mentoring; structured independent learning activities; and on-the-job, formal classroom, and online training. Senior staff train and spend time helping newer staff with both mastering technical issues and assimilating into the NRC culture. A major challenge is the multigenerational population now working together, each with different ways of learning and approaching work.

The NRC uses an integrated approach to learning to provide new employees with consistent information from branch to branch and division to division. To assist new employees, the NRC has developed a virtual orientation center. This advanced training tool allows new hires to enter a computer generated or virtual world where they can obtain information about the NRC organization, its mission, and employee benefits before starting their first day of work. Additionally, new hires receive position-specific training. The offices, such as the Office of Nuclear Reactor Regulation, have developed a qualification program that consists of three parts: general requirements, position-specific requirements, and oral qualification boards. The NRC is continuing to develop its qualification plans and other position-specific training for groups such as project engineers or project managers. It is also identifying course needs at its Technical Training Center and Professional Development Center.

Workforce Planning and Deployment. With a renewed emphasis on hiring to meet the expected increase in new reactor work, the Office of Nuclear Reactor Regulation realigned to emphasize the area of new reactors, and the Office of Nuclear Material Safety and Safeguards reorganized to enhance cooperation with States and to implement a holistic approach to fuel issues including transportation, storage, and disposal. The NRC's strategic workforce planning tool facilitated the changes in these two offices by allowing for a smoother planning process to improve workforce deployment, maintain technical capacity, and make informed decisions on human capital strategies for recruitment, development, and retention.

Leadership and Knowledge Management. The NRC has organized its leadership development programs into the Leaders Academy, consisting of systematic competency-based training, assessment, and development programs for all levels of leadership, from individual contributors to senior executives. The NRC also continues its executive succession planning process, through which it identifies skills needed and potential successors for senior leadership positions, determines development that would benefit executives to prepare them for such NRC positions, and considers strategies for filling positions for which NRC has few potential successors. The process informs selections for NRC positions and the establishment of executive development plans for all executives.

Knowledge management is a part of strategic human capital management, along with strategic workforce planning, recruitment, and training and development. As part of this effort, the NRC coordinates its activities to implement knowledge management strategies.

In addition, the NRC uses an agency-wide knowledge management plan that serves as a framework to integrate new and existing approaches that generate, capture, and transfer knowledge and information relevant to the NRC's mission. This plan includes both near- and long-term strategies, such as the following:

- Capture relevant critical knowledge of departing personnel
- Recapture departed knowledge where possible
- Communicate leadership's expectation for a knowledge-sharing culture
- Formalize knowledge management values and principles
- Incorporate knowledge management within process workflows

Current knowledge management and knowledge transfer activities include the following.

- Branch Chief and Team Leader Seminars - As the role of the NRC branch chiefs has evolved from the provisions of senior technical expertise to that of a manager, it is essential that the branch chiefs have the information they need to succeed in their positions. As a community of practice, the branch chiefs/team leaders meet monthly to hear presentations by agency experts in topics such as performance management, budget, and communications.

- Branch and Team Meetings - To ensure that staff members in each branch or team are kept up-to-date in areas under their purview, branch chiefs and team leaders hold regularly scheduled staff meetings. During some of these meetings, senior staff members give presentations to staff regarding an area in which they are considered experts or to pass their knowledge of past events on to newer staff. Some branch chiefs also have their more junior staff give presentations. This facilitates the interaction of junior staff with senior staff members, since the junior staff member may need to interview more senior staff to glean information for their presentations.

- Video Interviews – The NRC conducted a pilot project to capture knowledge from retiring senior staff using video interviews. One video captured knowledge regarding steam generators; another was entitled "Nuclear Knowledge for the Next Generation." The interviews included questions about licensing issues, recruiting and mentoring new hires, leadership, operations center experience, and reactor licensing performance metrics.

- Web Sites – The NRC has developed the "NRC Knowledge Center" Web page that links a number of communities and topics. Office-specific knowledge management programs supplement this agencywide site. For example the Office of Nuclear Reactor Regulation has a Web site devoted to knowledge management entitled "Sharing Expert Experience and Knowledge"; this site contains information such as the Inspector Best Practices Booklet and Inspector Newsletters, supervisor and team leader seminars, new employee orientation and training guide, key reference materials for reviews, qualification plans, strategic workforce planning, knowledge management, and other communities of practice.

Retaining Staff. The NRC is interested in retaining highly experienced staff who could retire if they wished, as well as more recent recruits whose skills are highly marketable outside the agency. The NRC relies on all aspects of its human resource management system, from providing challenging and meaningful work, comprehensive training and development, constructive performance management, and awards and recognition, to opportunities for career growth, financial incentives when needed, and a range of benefits, health, and work-life programs. These work-life programs include flexible work schedules and work-at-home plans. The agency's goal is to create an OCWE where people feel valued and challenged and in which employees and leaders at all levels model the NRC's core values: integrity, service, openness, commitment, cooperation, excellence, and respect.

8.1.6 Position of the NRC in the Governmental Structure

This section explains the relationship of the NRC to the executive branch, the States, and Congress.

8.1.6.1 Executive Branch

The components of the executive branch with which the NRC has the most frequent contact and interaction are the White House, OMB, U.S. Department of State, DOE, U.S. Environmental Protection Agency (EPA), FEMA, U.S. Department of Labor, U.S. Department of Transportation, and U.S. Department of Justice. Section 8.2 of this report discusses the NRC's relationship to DOE. The following summarizes the agency's relationships with other components of the Federal Government:

- The White House. As noted in Section 8.1.2.2, as an independent regulatory agency, the White House cannot directly set NRC policy. It can, however, influence NRC policy by (1) appointing Commissioners and Chairmen in whose outlook and judgment it has confidence and (2) making its views known on non-adjudicatory matters. In certain areas, such as national security policy, the Commission has declared its intent to give great weight to the views of the executive branch. In informal policy matters, such as rulemaking, White House and executive branch officials may properly try to influence NRC decisions, either publicly or privately. Ultimately, however, the NRC must make the decision and accept responsibility for it.

- U.S. Office of Management and Budget. The NRC submits its annual budget requests, including proposed personnel ceilings, to OMB for approval.

- U.S. Department of State. By law, the NRC must license the export and import of nuclear equipment and material. For significant applications, the Commission requests the U.S. Department of State to provide executive branch views on whether the license should be issued.

   The NRC also works with the U.S. Department of State negotiating international agreements in the nuclear field and interacting with IAEA and other international organizations of the United Nations, as well as NEA of the Organisation for Economic Co-operation and Development. In general, these interactions serve to develop policy on nuclear issues that are under NRC purview and to plan and coordinate programs of nuclear safety and safeguards assistance to other countries.

- U.S. Environmental Protection Agency. The responsibilities of the NRC and EPA intersect or overlap in areas in which EPA issues generally applicable environmental standards for activities that are also subject to NRC licensing. Examples include standards for high-level waste repositories, decommissioning standards, and standards for public and worker protection. EPA has the ultimate authority to establish generally applicable environmental standards to protect the environment from radioactive material.

- Federal Emergency Management Agency. FEMA assists the NRC's licensing process by preparing reviews and evaluations and by presenting witnesses to testify at licensing hearings. FEMA also participates with the NRC in observing and evaluating emergency exercises at nuclear plants. FEMA findings are not binding on the NRC, but they are presumed to be valid unless controverted by more persuasive evidence. FEMA is now part of the U.S. Department of Homeland Security (DHS).

- U.S. Department of Transportation. The NRC and the U.S. Department of Transportation share responsibility for the control of radioactive material transport. U.S. Department of Transportation regulations cover all aspects of transportation, including packaging, shipper and carrier responsibilities, documentation, and all levels of radioactive material.

- U.S. Department of Labor. The NRC monitors discrimination actions related to NRC-licensed activities filed with the U.S. Department of Labor under Section 211 of the Energy Reorganization Act and develops enforcement actions when there are properly supported findings of discrimination, either from the NRC's Office of Investigations or from U.S. Department of Labor adjudications.

- U.S. Department of Justice. The NRC has independent litigation authority. But any NRC litigation almost always requires coordination with the U.S. Department of Justice. Under the Administrative Orders Review Act (commonly called the Hobbs Act), the United States is a party to petitions for review challenging NRC licensing decisions or regulations.

The Office of Investigations, which investigates allegations of wrongdoing by NRC applicants and licensees, as well as by their contractors, normally works with the Fraud Section of the Criminal Division at the Department's Headquarters and with U.S. Attorneys.

The Office of the Inspector General reports to the Department of Justice whenever it has reasonable grounds to believe that an NRC employee or contractor has violated Federal law. The Inspector General refers cases for review for possible criminal prosecution to the U.S. Attorney's Office for the area in which the potential violation occurred. When the Department of Justice desires support from the Office of the Inspector General for investigations or grand jury work, it makes the request directly to the Inspector General.

8.1.6.2 The States (i.e., of the United States)

At the NRC, the Office of Federal and State Materials and Environmental Management Programs is responsible for establishing and maintaining effective communications and working relationships between the NRC and the States. This office serves as the primary contact for policy matters, keeping the States apprised about NRC activities and informing the NRC of State activities and views that may affect NRC policies, plans, and activities. Other NRC offices provide major support to implement State relations program policy and guidance, for example, through regional State liaisons and State agreements officers.

As explained in Article 7, the Atomic Energy Act confers on the NRC preemptive authority over health and safety regulation of nuclear energy and Atomic Energy Act materials. As a result, the general rule is that nuclear power plant safety, like airline safety, is the exclusive province of the Federal Government and cannot be regulated by the States. The courts would thus void a State law that attempted to set nuclear safety standards. However, the courts will not overturn a State law that regulates nuclear energy for purposes other than health and safety, such as economics, unless it conflicts with an NRC requirement. Similarly, the courts will not ordinarily question a State's declared purpose in enacting legislation.

However, the Atomic Energy Act did not entirely exclude States from the regulation of nuclear matters. Section 274 of the Act created the Agreement State Program, under which the NRC may relinquish its authority over most nuclear materials to those States willing to assume that authority. The NRC may not relinquish authority over such facilities as reactors, fuel reprocessing and enrichment plants, imports and exports, critical mass quantities of special nuclear material, high-level waste disposal, or certain other excepted areas.

Many States have signed formal agreements with the NRC and have assumed regulatory responsibility over certain byproduct, source, and small quantities of special nuclear materials. Agreement States receive no Federal funding to support their regulatory programs. The NRC conducts performance-based reviews of Agreement State programs to ensure that they remain adequate to protect public health and safety and are compatible with the NRC materials program.

Some States have shown a desire to participate in matters relating to nuclear power plants. In response, the NRC issued a policy statement in February 1989 declaring its intent to cooperate with States in the area of nuclear power plant safety by keeping States informed of matters of interest to them and considering proposals for State officials to participate in NRC inspection activities, pursuant to a memorandum of understanding between the State and NRC. The policy statement makes clear that States must channel their contacts with the NRC through a single State Liaison Officer, appointed by the Governor. States are authorized only to observe and assist in NRC inspections of reactors, and they cannot conduct their own independent health and safety inspections.

Through its intergovernmental liaison program, the NRC works in cooperation with Federal, State, and local governments; interstate organizations; and Native American Tribal Governments to maintain effective relations and communications with these organizations and to promote greater awareness and mutual understanding of the policies, activities, and concerns of all parties involved as they relate to radiological safety at NRC-licensed facilities.

8.1.6.3 Congress

The following oversight committees and subcommittees in the Senate and House have jurisdiction over aspects of the NRC's activities:

- Senate Oversight.   In the U.S. Senate, the Committee on the Environment and Public Works has jurisdiction over domestic nuclear regulatory activities.   Within the committee, the Subcommittee on Clean Air and Nuclear Safety has responsibility for regulation and oversight of the NRC.   The Energy and Natural Resources Committee and the Environment and Public Works Committee share jurisdiction over nuclear waste issues.

- House Oversight.   In the U.S. House of Representatives, the Committee on Energy and Commerce has jurisdiction over domestic nuclear regulatory activities.   Within the committee, the Subcommittee on Energy and Environment has responsibility for regulation and oversight of the NRC.

- Other Relevant Committees.   In addition to the committees and subcommittees mentioned above, the House and Senate Appropriations Subcommittees on Energy and Water Development play a key role in approving the Commission's annual budget.   A number of other committees frequently interface with the NRC concerning international affairs, research, security, and general Governmental operations.

8.1.7 Report of the Integrated Regulatory Review Service Self-Assessment Team

The U.S. has invited an IAEA IRRS Mission scheduled for October 17 – 29, 2010.   The preparatory meeting took place October 21 – 23, 2009.   Subsequent to this meeting, a new IRRS Mission team leader was assigned.   Therefore, a second preparatory meeting with the new team leader took place on March 12, 2010.   To prepare for the mission, the NRC performed a complementary self-assessment in 2009 to update a self-assessment previously performed in 2007.

The mission will focus specifically on the operating power reactor program.   U.S. preparatory activities initially followed the IAEA procedure titled, "Guidelines for the Integrated Regulatory Review Service (IRRS)," dated February 2008, but were realigned to follow the February 2010 IRRS guidance following its issuance.   The U.S. mission will include all 10 core modules of the 2010 guidance, as well as some additional thematic and optional modules, and will discuss three Elective Policy Issues.   The three Policy Issues are: (1) transparency and openness, (2) long-term operation and aging management of nuclear facilities, and (3) human resources and knowledge management.

8.2 Separation of Functions of the Regulatory Body from Those of Bodies Promoting Nuclear Energy

Although both the NRC and DOE have responsibilities for managing nuclear facilities and materials, they maintain separate, independent functions.   The partitioning of the U.S. Atomic Energy Commission in the mid-1970s provided distinct entities for the U.S. Government's regulatory and promotional responsibilities in nuclear applications.

Specifically, the Energy Reorganization Act redistributed the functions performed by the U.S. Atomic Energy Commission to two new agencies. This act created the NRC to regulate the commercial nuclear power sector and ERDA to promote energy and nuclear power development and to develop defense applications. The NRC was established as an independent authority to regulate the possession and use of nuclear materials as well as the siting, construction, and operation of nuclear facilities. ERDA was established to ensure the development of all energy sources, increase the efficiency and reliability of energy resource use, and carry out the other functions, including but not limited to the U.S. Atomic Energy Commission military and production activities and general basic research activities.

The NRC performed its regulatory mission by issuing regulations, licensing commercial nuclear reactor construction and operation, licensing the possession of and use of nuclear materials and wastes, safeguarding nuclear materials and facilities from theft and radiological sabotage, inspecting nuclear facilities, and enforcing regulations. The NRC regulates the commercial nuclear fuel cycle materials and facilities. Regarding the regulatory control of commercial spent nuclear fuel and radioactive waste, the NRC is responsible for licensing commercial nuclear waste management facilities, independent spent fuel management facilities, and DOE facilities for the disposal of high-level waste and spent fuel.

DOE addresses the U.S. Government's need to unify energy organization and planning. The DOE Organization Act brought a number of Federal agencies and programs, including ERDA, into a single agency with responsibilities for nuclear energy technology and nuclear weapons programs. Over the past decade, DOE has expanded its new nuclear-related activities to include nonproliferation and the environmental cleanup of contaminated sites and facilities. DOE retains authority under the Atomic Energy Act for regulating its nuclear activities, including the responsibility for activities such as regulating the disposal of its own low-level radioactive waste.

ARTICLE 9.   RESPONSIBILITY OF THE LICENSE HOLDER

Each Contracting Party shall ensure that prime responsibility for the safety of a nuclear installation rests with the holder of the relevant license and shall take the appropriate steps to ensure that each such license holder meets its responsibility.

The NRC, through the Atomic Energy Act, ensures that the prime responsibility for the safety of a nuclear installation rests with the licensee.   Steps the NRC takes to ensure that each licensee meets its primary responsibility include the licensing process, discussed in Articles 18 and 19, the Reactor Oversight Process, discussed in Article 6, and the enforcement program, discussed below.   This update revises the debt collection dollar amount and discusses the Alternative Dispute Resolution Program and current experience.

9.1   Introduction

The NRC's regulatory programs continue to be based on the premise that the safety of commercial nuclear power reactor operations is the responsibility of NRC licensees.   The NRC is responsible for regulatory oversight of licensee activities to ensure that safety is maintained. The NRC reviews the safety of a reactor design and the capability of an applicant to design, construct, and operate a facility.   If an applicant satisfies the requirements of the *Code of Federal Regulations*, the NRC then issues a license to operate the facility.   Such licenses specify the terms and conditions of operation to which a licensee must conform.   Failure to conform subjects the licensee to enforcement action, which can include modifying, suspending, or revoking the license.   The NRC can also order particular corrective actions or issue civil penalties.   The following sections discuss these enforcement mechanisms in greater detail.

9.2   The Licensee's Prime Responsibility for Safety

As discussed in Article 7 of this report, the Atomic Energy Act, Section 103, Chapter 10, grants the NRC authority to issue licenses for nuclear reactor facilities.   Moreover, Section 103 states that these licenses are subject to such conditions as the NRC may establish by rule or regulation to implement the purposes and provisions of the Atomic Energy Act.   Consistent with the Act, before issuing a license, the Commission determines that the applicant is (1) equipped and agrees to observe such safety standards to protect health and minimize danger to life or property as the Commission may establish by rule and (2) agrees to make available to the Commission such technical information and data about activities under such license as the Commission may determine necessary to promote the common defense and security and to protect public health and safety.

Embedded in each license is the explicit responsibility for the license holder to comply with the terms and conditions of the license and the applicable Commission rules and regulations.   The licensee is ultimately responsible for the safety of its activities and the safeguarding of nuclear facilities and materials used in operation.

When the Commission or licensee determines that the licensee is not complying with the Commission's rules or regulations, the NRC takes action to ensure that the facility is returned to a condition compliant with its license.

9.3 NRC Enforcement Program

As discussed in Article 7, the NRC has enforcement powers.   As discussed in Sections 7.2.3 and 7.2.4, the enforcement process complements the Reactor Oversight Process.   The NRC uses enforcement as a deterrent to emphasize the importance of compliance with regulatory requirements and to encourage prompt identification and prompt, comprehensive correction of violations.

The NRC identifies violations through inspections and investigations.   All violations are subject to civil enforcement action and may be subject to criminal prosecution.   Unlike the burden of proof standard for criminal actions (beyond a reasonable doubt), the NRC uses the Administrative Procedure Act standard (preponderance of evidence) in enforcement proceedings. After an apparent violation is identified, it is assessed in accordance with the Commission's enforcement policy, described in the NRC Enforcement Policy, last updated on November 28, 2008, which is available to NRC licensees and members of the public.   The NRC Office of Enforcement maintains the current policy statement on the NRC's public Web site. Because it is a policy statement and not a regulation, the Commission may deviate from it, as appropriate, given the circumstances of a particular case.

The NRC has three primary enforcement sanctions available: notices of violation, civil penalties, and orders.[3]   A notice of violation identifies a requirement and how it was violated, formalizes a violation pursuant to 10 CFR 2.201, "Notice of Violation," requires corrective action, and normally requires a written response.   A civil penalty is a monetary fine issued under authority of the Atomic Energy Act, Section 234, or the Energy Reorganization Act, Section 206.   Section 234 of the Atomic Energy Act provides for penalties of up to $100,000 per violation per day; however, that amount is adjusted every 4 years by the Federal Civil Penalties Inflation Adjustment Act of 1990, as amended by the Debt Collection Improvement Act of 1996, and is currently $140,000. Section 161 of the Atomic Energy Act gives the Commission broad authority to issue orders; this authority extends to any area of licensed activity that affects public health and safety or the common defense and security.   Orders modify, suspend, or revoke licenses, or they may require specific actions by licensees or persons.   The NRC issues notices of violations and civil penalties on the basis of violations.   The agency may issue orders for violations or, in the absence of a violation, because of a concern involving public health and safety or the common defense and security.

After identifying a violation, the NRC assesses its significance by considering the following factors:

- actual safety consequences
- potential safety consequences
- potential for impacting the NRC's ability to perform its regulatory function
- any willful aspects of the violation

---

[3]   The NRC also uses administrative actions, such as notices of deviation, notices of nonconformance, confirmatory action letters, and demands for information to supplement its enforcement program.

Given those factors, the NRC takes one of the following actions based on the significance of the violation:

- assigns a severity level, ranging from Severity Level IV (more than minor concern) to Severity Level I (the most significant)

- associates the violation with findings assessed through the Reactor Oversight Process significance determination process (described in Article 6) and assigns a color code of green, white, yellow, or red based on increasing risk significance

The Commission recognizes that there are violations of minor safety or environmental concern that are below Severity Level IV violations, as well as below violations associated with green findings. These minor violations are not assigned a severity level category or a color assessment.

The NRC may hold a pre-decisional enforcement conference or a regulatory conference with a licensee before making an enforcement decision if (1) escalated enforcement action appears warranted, (2) the NRC decides a conference is necessary, or (3) the licensee requests it. The purpose of the conference is to obtain information to assist the NRC in determining the appropriate enforcement action, such as a common understanding of facts, root causes, and missed opportunities associated with the apparent violations; corrective actions taken or planned; and the significance of issues and the need for lasting, comprehensive corrective actions.

At several junctions during the enforcement process involving cases of discrimination or willful violation of NRC regulations, the agency offers its licensees (including their contractors) or individuals the opportunity to participate in the Alternative Dispute Resolution Program. Alternative dispute resolution is a general term encompassing various techniques for resolving conflict outside of court using a neutral third party. The NRC uses mediation, a technique in which a neutral mediator with no decisionmaking authority helps parties clarify issues, explore settlement options, and evaluate how best to advance their respective interests. Neutral mediators are selected from a roster of experienced mediators provided by a neutral program administrator who is under contract with the NRC. The mediator's responsibility is to assist the parties in reaching an agreement. However, the mediator has no authority to impose a resolution upon the parties. Mediation is a confidential and voluntary process. If the parties to the process (the NRC and the licensee or individual) agree to use alternative dispute resolution, they select a mutually agreeable neutral mediator and share equally the cost of the mediator's services. In cases in which the NRC and the other party reach an agreement, the agency issues a confirmatory order reflecting the terms of the agreement.

The agency normally assesses civil penalties for Severity Level I and II violations, as well as knowing and conscious violations of the reporting requirements of Section 206 of the Energy Reorganization Act. Civil penalties are considered for Severity Level III violations. Although not normally used for violations associated with the Reactor Oversight Process, civil penalties (and the use of severity levels) are considered for issues that are willful, that have the potential to affect the regulatory process, or that have actual consequences.

Although each severity level may have several associated considerations, the outcome of the assessment process for each violation or problem (absent the exercise of discretion) results in one of three outcomes, which may involve no civil penalty, a base civil penalty, or twice the base civil penalty.

The NRC may issue orders to modify, suspend, or revoke a license; issue orders to cease and desist from a given practice or activity; or take such other action as may be proper. The agency may issue orders in lieu of, or in addition to, civil penalties. Additionally, the NRC may issue an order to impose a civil penalty when a licensee refuses to pay a civil penalty or an order to an unlicensed person (including vendors) when the agency has identified deliberate misconduct. By statute, a licensee or individual may request a hearing upon receiving an order. Orders are normally effective after a licensee or individual has had an opportunity to request a hearing (i.e., 30 days). However, orders can be made immediately effective without prior opportunity for a hearing when the agency determines it is the best interest of public health and safety to do so. Subsequent to the hearing process, a licensee or individual may appeal the administrative hearing decision to the Commission and, if desired, appeal the Commission's decision to a U.S. court of appeals.

Providing interested stakeholders with enforcement information is very important to the NRC. Conferences that are open to public observation appear in the listing of public meetings on the NRC's public Web site. The agency issues a press release for each proposed civil penalty or order. All orders are published in the *Federal Register*. Significant enforcement actions (including actions to individuals) are included in the enforcement document collection in the Electronic Reading Room of the NRC's public Web site.

During 2008, the NRC issued a variety of significant enforcement actions to operating power reactors. These actions included 23 escalated notices of violation without civil penalties, 3 civil penalties, and 3 orders.

During 2009, the NRC issued a variety of significant enforcement actions to operating power reactors including 22 escalated notices of violation without civil penalties, 1 civil penalty, and 4 orders.

To provide accurate and timely information to all interested stakeholders and enhance the public's understanding of the enforcement program, the NRC publishes related information on the agency's public Web site.

ARTICLE 10.   PRIORITY TO SAFETY

Each Contracting Party shall take the appropriate steps to ensure that all organizations engaged in activities directly related to nuclear installations shall establish policies that give due priority to nuclear safety.

NRC policies that give due priority to safety covered under this article are PRA policy statements and policies that apply to licensee safety culture and safety culture at the NRC.

Other articles (e.g., Articles 6, 14, 18, and 19) also discuss activities undertaken to achieve nuclear safety at nuclear installations.

Updates to this section discuss new regulations, developments in PRA, and safety culture.

10.1   Background

The United States has made substantial progress in developing and using the results of PRAs for all operating reactor facilities, and the NRC has developed extensive guidance regarding the role of PRA in U.S. regulatory programs.   The NRC has extensively applied information gained from PRA to complement other engineering analyses in improving issue-specific safety regulation and in changing the current licensing bases for individual plants.   The move toward risk-informing the current regulations and processes continues to mark perhaps the most significant changes at the NRC.   For example, 10 CFR 50.69, "Risk-Informed Categorization and Treatment of Structures, Systems, and Components," modifies the scope of the special treatment regulations by creating an alternative regulatory framework that enables licensees to use a risk-informed approach to categorize structures, systems, and components (SSCs), and their associated treatment, according to their safety significance.   As another example, 10 CFR 50.48(c) allows an operating nuclear power plant licensee to adopt a risk-informed, performance-based fire protection program (discussed further in the Survey of Regulatory and Current Issues section of this report).   The NRC is continuing a program to develop additional changes to the specific technical requirements in the body of 10 CFR Part 50.

10.2   Probabilistic Risk Assessment Policy

Three policy statements form the basis for the NRC's current treatment of PRA and the related regulatory safety goals and objectives - the "Policy Statement on Severe Reactor Accidents Regarding Future Designs and Existing Plants," dated August 8, 1985; the "Safety Goals for the Operation of Nuclear Power Plants; Policy Statement; Republication," dated August 21, 1986; and the "Policy Statement on Use of PRA Methods in Nuclear Activities," dated August 16, 1995. Previous U.S. National Reports have detailed these policies.

10.3   Applications of Probabilistic Risk Assessment

The NRC applies PRA to resolve severe accident issues, evaluate new and existing requirements and programs, implement risk-informed regulation, and improve data and methods of risk analysis.   The NRC also engages in cooperative activities with industry (such as pilot programs for 10 CFR 50.69, 10 CFR 50.48(c), and RG 1.200, Revision 2, "An Approach for Determining the Technical Adequacy of Probabilistic Risk Assessment Results for Risk-Informed Activities," dated March 2009) and in activities that assess risk in determining plant-specific changes to the licensing basis. The NRC staff will use RG 1.200 to assess technical adequacy of the supporting PRA for all risk-informed applications.

The NRC maintains a risk-informed and performance-based plan, updated annually, which sets forth the agency's planned actions to make its regulatory activities risk informed and performance based. In the past, the Risk-Informed Regulation Implementation Plan (for example, SECY-09-0159 "Annual Update of the Risk-Informed and Performance-Based Plan," dated October 27, 2009) focused largely on risk-informed initiatives. The current improved plan has expanded the objectives to more fully achieve a risk-informed and performance-based regulatory structure. The NRC has created a public Web site for the risk-informed and performance-based plan with links to documents that specifically describe activities and status.

The NRC and industry representatives have cooperated in a number of activities and pilot programs to develop and apply risk-informed methodologies for specific regulatory applications. The staff uses the lessons learned from these activities to enhance the effectiveness of developed guidance. These activities, described in the sections below, include special treatment, inservice inspection, technical specification changes, and standards development.

10.3.1   Risk-Informed Special Treatment

The agency has approved or is reviewing several applications of risk-informed inservice testing, of generally limited scope. For example, in August 2001, the staff granted a risk-informed exemption request from the licensee of the South Texas Project regarding special treatment requirements for low-risk and nonrisk-significant safety-related nuclear components (including an exemption from prescriptive inservice testing requirements). Having successfully implemented this exemption, the staff developed a new rule, 10 CFR 50.69 (discussed in Section 10.1 of this report), to allow the application of risk insights to reduce the special treatment requirements in 10 CFR Part 50 for SSCs that are categorized as being of low safety significance.

The Commission approved the final rule, with some modifications, in October 2004. The final rule was published in the *Federal Register* on November 22, 2004. The NRC staff issued RG 1.201, Revision 1, "Guidelines for Categorizing Structures, Systems, and Components in Nuclear Power Plants According to Their Safety Significance," in May 2006.

A topical report, WCAP-16308-NP, Revision 0, "Pressurized Water Reactor Owners Group 10 CFR 50.69 Pilot Program – Categorization Process – Wolf Creek Generating Station," dated September 25, 2006, proposed a categorization process used by Wolf Creek Nuclear Operating Corporation in support of a future licensee submittal requesting approval to implement 10 CFR 50.69. The staff completed its review of the topical report and issued its final safety evaluation in March 2009. The staff found the categorization process described in the topical report to be acceptable, but it did not approve or endorse any specific treatment process. Treatment programs being implemented under 10 CFR 50.69 do not require prior approval from the NRC as part of the license amendment review process.

The staff plans to develop guidance for sample inspections to be conducted at plants voluntarily choosing to implement 10 CFR 50.69. The performance of sample inspections is consistent with the statement of considerations accompanying the final 10 CFR 50.69 rule. The staff plans to issue draft guidance to obtain stakeholder input and issue final guidance by summer 2011. Inspection efforts will be focused on the most risk significant aspects related to implementation of 10 CFR 50.69 (i.e., proper categorization of SSCs and treatment of Risk-Informed Safety Class (RISC)-1 and RISC-2 SSCs). Additionally, the inspections are expected to be performance based, with SSCs with a lower safety significant function, such as those classified RISC-3, not receiving a major portion of inspection focus unless adverse performance trends are observed.

The staff recognizes the need for an effective, stable, and predictable regulatory climate for the implementation of 10 CFR 50.69. Inspection guidance developed with industry stakeholder input is viewed as an efficient vehicle for reaching a common understanding of what constitutes an acceptable treatment program for SSCs, since the NRC does not review specific treatment plans as part of a licensee's application to implement 10 CFR 50.69.

10.3.2 Risk-Informed Inservice Inspection

The NRC uses the inservice inspection guidance in RG 1.178, Revision 1, "An Approach for Plant Specific Risk-Informed Decision-making for Inservice Inspection of Piping," dated September 2003, and NUREG-0800, "Standard Review Plan for the Review of Safety Analysis Reports for Nuclear Power Plants: LWR Edition," Section 3.9.8, "Risk-Informed Inservice Inspection of Piping," dated September 2003. The agency-approved industry methodologies, one developed by the Westinghouse Owners Group and the other by EPRI, regarding alternatives to the American Society of Mechanical Engineers (ASME) Boiler and Pressure Vessel Code (ASME Code), Section XI, Inservice Inspection Program continue to be used for inservice inspections.

ASME has developed Code Case N-716, "Alternative Piping Classification and Examination Requirements, Section XI Division 1." Code Case N-716 is founded, in large part, on the risk-informed inservice inspection process as described in EPRI Topical Report 112657, Revision B-A, "Revised Risk-Informed Inservice Inspection Evaluation Procedure," dated December 1999, which the NRC reviewed and approved. Code Cases provide alternatives to existing ASME Code requirements that ASME has developed and approved. RG 1.147, Revision 15, "Inservice Inspection Code Case Acceptability, ASME Section XI, Division 1," dated October 2007, identifies the Code Cases that the NRC has determined to be acceptable alternatives to applicable parts of ASME Code, Section XI. RG 1.147 has not endorsed Code Case N-716 because the technical adequacy of a PRA that can be used to develop a risk-informed inservice inspection program is not well defined. The NRC has reviewed and approved about 12 plant-specific risk-informed inservice inspection programs that are based on the methodology described in Code Case N-716 supplemented with information related to the plant's PRA. By letter dated February 18, 2009, EPRI submitted for NRC staff review Topical Report 1018427, "Nondestructive Evaluation: Probabilistic Risk Assessment Technical Adequacy Guidance for Risk-Informed In-Service Inspection Programs." The staff is scheduled to complete its review of Topical Report 1018427 by December 2010. If the NRC endorses Topical Report 1018427, it will determine whether RG 1.147 can endorse Code Case N-716, supported by Topical Report 1018427. Licensees may implement Code Cases endorsed in RG 1.147 without prior NRC staff review and approval.

The NRC regularly participates in the ASME Code development process to resolve issues regarding risk-informed inservice inspection methodology.

10.3.3 Risk-Informed Technical Specification Changes

Since the mid-1980s, the NRC has reviewed and granted improvements to technical specifications that are based, at least in part, on PRA insights. In its "Final Policy Statement on Technical Specification Improvements for Nuclear Power Reactors," published in the *Federal Register* on July 22, 1993, the Commission stated that it expects licensees to use a plant-specific

PRA or risk survey in preparing submittals related to technical specifications. The Commission reiterated this point when it revised 10 CFR 50.36, "Technical Specifications," in July 1995.

The NRC continues to use RG 1.177, "An Approach for Plant-Specific, Risk-Informed Decisionmaking: Technical Specifications," dated August 1998, and a companion section of NUREG-0800 to guide licensees in making risk-informed changes to plant technical specifications. The agency uses RG 1.177 as well as RG 1.174, Revision 1, "An Approach for Using Probabilistic Risk Assessment in Risk-Informed Decisions on Plant-Specific Changes to the Licensing Basis," dated November 2002, to improve plant technical specifications. The industry and the NRC continue to increase the use of PRA in developing improvements to technical specifications. As discussed in a letter from NEI to the NRC dated June 8, 2001 (Agencywide Documents Access and Management System Accession No. ML011690233), the industry developed eight separate initiatives to improve existing technical specification configuration control requirements through use of risk insights. The following summarizes the major accomplishments in this area:

- Initiative 1, "Modified End States" - This initiative would allow (following a risk assessment) some equipment to be repaired during hot shutdown rather than cold shutdown. The NRC has approved the topical reports and model applications supporting this initiative for BWRs and for Combustion Engineering and Babcock & Wilcox plants. The staff is currently reviewing the Westinghouse topical report, submitted September 2005.

- Initiative 4b, "Risk-Informed Completion Times" - The overall objective of this initiative is to modify technical specifications to reflect a configuration risk management approach that is more consistent with the approach of the Maintenance Rule (10 CFR 50.65(a)(4)). Industry guidance has been approved, and the South Texas Project pilot was approved in 2007. The NRC expects to receive a model application in 2010 for review and approval.

- Initiative 5b, "Risk Informed Method for Control of Surveillance Frequencies" - This initiative allows licensees to modify the frequency of technical specification surveillances based on test data and a risk-informed evaluation. The staff approved industry guidance and a model application, and it has approved pilot applications for the Limerick Generating Station in 2006 and Diablo Canyon in 2009. The staff is currently receiving and reviewing applications for this initiative.

- Initiative 6, "Modification of Limiting Condition for Operation 3.0.3, Actions and Completion Times" - This initiative provides a 24-hour completion time for a limited scope of technical specification systems when both safety trains are inoperable. The industry is in the process of resolving discrepancies between its Combustion Engineering topical reports WCAP-16125, Revisions 1 and 2, "Justification for Risk-Informed Modifications to Selected Technical Specifications for Conditions Leading to Exigent Plant Shutdown," dated December 2007 and May 2009, respectively, and the NRC's draft safety evaluation. The NRC is currently reviewing the May 2009 document. The NRC expects to receive a BWR topical report in FY2010.

- Initiative 7, "Non-Technical Specifications Support System Impact in Technical Specifications System Operability": This initiative permits a risk-informed delay time before entering limiting condition for operation actions for inoperability attributable to a loss of support function provided by equipment not addressed in technical specifications. Guidance documents have been approved for snubbers and hazard barriers, and the industry is considering additional proposals.

- Initiative 8, "Remove/Relocate Non-Safety and Non-Risk Significant Systems from Technical Specifications" - This initiative would review technical specifications to remove certain system functions that had been included solely because they were judged to be risk significant at one time, but additional analysis could show them not to be. The industry and staff are in preliminary discussions on this initiative.

10.3.4 Development of Standards

The NRC worked with ASME and the American Nuclear Society (ANS) to develop a national consensus standard for PRA quality. In February 2009, ASME and ANS issued their joint PRA quality standard, ASME/ANS-RA-Sa-2009, "Standard for Level 1/Large Early Release Frequency Probabilistic Risk Assessment for Nuclear Power Plant Applications," and the NRC endorsed it in RG 1.200, Revision 2, in March 2009. The Survey of Current Regulatory and Safety Issues section of this report provides further information on the PRA standard for external events.

The agency plans further revisions to the RGs to incorporate revisions to the ASME/ANS standards as they are published, including standards addressing low power and shutdown modes, and Level 2 and 3 PRA.

10.4 Safety Culture

An important means to implement any policy that gives due priority to safety is to foster a strong safety culture in the organization. The following discussion focuses upon safety culture, and efforts to improve safety culture, in the NRC and in the nuclear industry.

10.4.1 NRC Monitoring of Licensee Safety Culture

This section covers the policies, programs, and practices that apply to licensee safety culture.

10.4.1.1 Background

Section 6.3.2 of this report describes the Reactor Oversight Process. Based on lessons learned from the Davis-Besse reactor pressure vessel head degradation event and other considerations, the NRC enhanced the Reactor Oversight Process to more fully address safety culture and identify safety culture problems earlier so that corrective steps can be taken to address the problems and prevent further plant performance degradation.

10.4.1.2 Enhanced Reactor Oversight Process

The NRC has adopted the IAEA International Nuclear Safety Advisory Group's definition of safety culture provided in Safety Series No.75-INSAG-4, "Safety Culture," dated February 1991, as "that assembly of characteristics and attitudes in organizations and individuals which establishes that, as an overriding priority, nuclear safety issues receive the attention warranted by their significance."

On the basis of a review of safety culture attributes developed or applied by IAEA, NEA, INPO, regulatory bodies in other countries, and other domestic organizations, staff expertise, and input and feedback from NRC stakeholders, the staff identified the following components as important to safety culture:

- decisionmaking
- resources
- work control
- work practices
- corrective action program
- operating experience
- self- and independent assessments
- environment for raising safety concerns
- preventing, detecting, and mitigating perceptions of retaliation
- accountability
- continuous learning environment
- organizational change management
- safety policies

The Reactor Oversight Process inspection guidance documents define each one of the safety culture components in a greater level of detail (e.g., cross-cutting aspects). The Reactor Oversight Process applies the safety culture components, and their associated aspects, in different ways. The first nine safety culture components are applied in the baseline inspection and assessment program. All 13 safety culture components are applied in selected baseline, event followup, and supplemental inspection procedures (IPs).

Licensees perform periodic, voluntary self-assessments of safety culture in accordance with industry guidelines (further discussed in Section 3 of this report). There are no regulatory requirements for licensees to perform safety culture assessments routinely. However, depending on the extent of deterioration of licensee performance, the NRC has a range of expectations regarding regulatory actions and licensee safety culture assessments, as described below.

The Reactor Oversight Process employs a graded approach, such that plants that are performing in a specified manner warrant only a routine level of inspection and oversight. However, as licensee performance deteriorates, inspection and oversight become increasingly more intrusive to ensure safe plant operation. The Reactor Oversight Process safety culture enhancements continue to allow licensees to self-diagnose and implement corrective actions for their performance problems before the NRC performs followup inspections.

For most licensees (i.e., those listed in the Licensee Response column, Column 1, of the Reactor Oversight Process Action Matrix), the NRC performs the baseline inspection program. In the routine or baseline inspection program, the inspector will develop an inspection finding and then identify whether an aspect of a safety culture component is a significant causal factor of the finding. The NRC communicates the inspection findings to the licensee along with the associated safety culture aspect.

The NRC revised the IP that focuses on problem identification and resolution to allow inspectors to have the option to review licensee self-assessments of safety culture. The problem identification and resolution IP also instructs inspectors to be aware of safety culture components when selecting samples. In addition, questions related to safety-conscious work environment were enhanced in the procedure.

The agency revised IP 71153, "Followup of Events and Notices of Enforcement Discretion," dated June 10, 2006, to direct inspection teams to consider contributing causes related to the safety culture components as part of their efforts to fully understand the circumstances surrounding an event and its probable cause(s).

As part of the assessment process (conducted twice per year), the NRC considers the aspects of safety culture components associated with inspection findings to determine whether common themes exist at a plant. If, over three consecutive assessment periods (i.e., 18 months), a licensee has the same safety culture issue with the same common theme, the NRC may ask the licensee to conduct a safety culture self-assessment.

As licensee performance declines (Regulatory Response column, Column 2, of the Reactor Oversight Process Action Matrix), the inspectors, through a specific supplemental IP, verify that the licensee's root cause, extent of condition, and extent of cause evaluations for the risk-significant finding(s) appropriately considered the safety culture components.

When the licensee performance degrades further (Degraded Cornerstone column, Column 3, of the Reactor Oversight Process Action Matrix), the NRC expects that the licensee's root cause evaluation for the risk-significant finding(s) determined whether any safety culture component contributed to the risk-significant performance issues. If through the conduct of supplemental IP 95002, "Inspection for One Degraded Cornerstone or any Three White Inputs in a Strategic Performance Area", dated June 22, 2006, the NRC determines that the licensee did not recognize that safety culture components caused or significantly contributed to the risk-significant performance issues, the NRC may request the licensee to complete an independent assessment of its safety culture.

Finally, for licensees with more significant performance degradation (Multiple/Degraded Cornerstone column, Column 4, of the Reactor Oversight Process Action Matrix), the NRC will expect the licensee to conduct a third-party independent assessment of its safety culture. The NRC will review the licensee's assessment and will conduct an independent assessment of the licensee's safety culture via a specific supplemental IP that was substantially revised to provide guidance for these assessments. The staff applied this revised IP for the first time at the Palo Verde plant in 2007.

In July 2006, the NRC implemented revisions to the Reactor Oversight Process inspection and assessment processes related to safety culture. The NRC inspectors received training on safety culture in general and on the changes to the Reactor Oversight Process before implementation. Ongoing inspector training now includes safety culture topics. In 2008, the NRC conducted a self-assessment to review the changes to the Reactor Oversight Process over the initial 18-month implementation period. Lessons learned from the initial 18-month implementation period and from the Palo Verde supplemental inspection resulted in IP and program guidance changes. Some of the more significant changes included using a graded approach to evaluating safety culture assessments and the inclusion of additional guidance related to safety-conscious work environment considerations.

The safety culture changes made to the Reactor Oversight Process were intended to provide the NRC staff with (1) better opportunities to consider safety culture weaknesses and to encourage licensees to take appropriate actions before significant performance degradation occurs, (2) a process to determine the need to specifically evaluate a licensee's safety culture after performance problems have resulted in the placement of a licensee in the Degraded Cornerstone column of the Reactor Oversight Process Action Matrix, and (3) a structured process to evaluate the licensee's safety culture assessment and to independently conduct a safety culture assessment for a licensee in the Multiple/Repetitive Degraded Cornerstone column of the action matrix.

By using the existing Reactor Oversight Process framework, the NRC's safety culture oversight activities are based on a graded approach and remain transparent, understandable, objective, risk informed, performance based, and predictable.

10.4.2 The NRC Safety Culture

As previously noted in Section 10.4.1, the NRC recognizes the importance of nuclear plant operators establishing and maintaining a strong safety culture -- a work environment where management and employees are dedicated to putting safety first.   In November 2009, the agency published the draft Safety Culture Policy Statement in the *Federal Register* that set forth the expectation that all licensees and certificate holders establish and maintain a positive safety culture.   Similarly, given the NRC's safety and security mission, the NRC recognizes the importance of maintaining its own strong safety culture and the need to continuously seek to improve its internal organizational effectiveness.

In response to the identification of licensee safety culture weaknesses as contributing factors to events, the agency revised the Reactor Oversight Process in 2006 to better address safety culture; enhancement efforts to the Reactor Oversight Process continue.   These external efforts prompted internal reflection on how to improve the agency's own safety culture.   Accordingly, in October 2008, the agency chartered the NRC Internal Safety Culture Task Force to provide a report to the Commission outlining potential initiatives that could improve the agency's internal safety culture.

Based on the results from a range of data collection activities and the experience and knowledge of its members, the NRC Internal Safety Culture Task Force developed a set of recommendations.   These recommendations, which are being implemented, aim to create effective and lasting improvements for supporting a strong safety culture.   Actions include the following:

- the appointment of an agency Safety Culture Program Manager
- integrating safety culture into the NRC's Strategic Plan and integrating performance management tools
- developing training on safety culture principles and expectations
- evaluating the agency's problem identification, evaluation, and resolution processes
- establishing clear expectations and accountability for maintaining current policies and procedures

SECY-09-0068, "Report of the Task Force on Internal Safety Culture," dated April 27, 2009, and SECY-10-0009, "Internal Safety Culture Update," dated January 26, 2010, provide more details, including, in the latter, a status on the implementation of the recommendations in the task force report.

Complementing this new initiative is the agency's ongoing effort to encourage the free and open discussion of differing professional views in order to develop sound regulatory policy and decisions. The NRC strives to establish and maintain an OCWE that encourages all employees and contractors to promptly voice differing views without fear of retaliation. The staff created the OCWE Web page (http://www.nrc.gov/about-nrc/values/open-work-environment.html) in 2007 to clearly communicate that the NRC encourages trust, respect, and open communication to foster and promote a positive work environment that maximizes the potential of all individuals and improves regulatory decisionmaking. The OCWE Web page also identifies some of the policies in place that permit employees at all levels in all areas to provide professional views on virtually all matters pertaining to the agency's mission.

The NRC Open Door Policy (first communicated to agency employees in 1976), the NRC Differing Professional Opinions Program (formally established in 1980), and the NRC Non-Concurrence Process (established in 2006) illustrate the NRC's commitment to the free and open discussion of professional views. In 2008, the NRC created the NRC Team Player awards, which recognize and celebrate behaviors that support an OCWE where differing views are welcomed, valued, fairly considered, and addressed.

The agency uses the Office of the Inspector General's periodic Safety Culture and Climate Survey as a means to assess the effectiveness of these new and existing safety culture efforts. In 1998, the Office of the Inspector General conducted the first in a continuing series of Safety Culture and Climate Surveys as a means to identify areas for additional organizational improvements. The surveys are voluntary, provide for anonymity, and are offered to all NRC employees, supervisors, and managers. In addition, the use of a survey makes it possible to compare category-level results for the NRC to other U.S. organizations that have completed such a survey. The Office of the Inspector General has conducted the Safety Culture and Climate Surveys four times: 1998, 2002, 2005, and most recently in 2009.

An unprecedented 87-percent survey response rate in 2009 surpassed the response rate of 71-percent in 2005 and the average rate of return of 80 percent of high-performance companies. Compared to results for the 2005 Safety Culture and Climate Survey, the agency saw substantial improvements in 16 of 17 categories surveyed, and scores were generally in line with or better than those of U.S. high-performance companies. Those categories showing outstanding improvement include the following:

- mission and Strategic Plan
- image
- performance management
- commitment to continuous improvement
- management leadership
- OCWE

The Office of the Inspector General's detailed report on the 2009 survey is available on the NRC's Web site at http://www.nrc.gov/reading-rm/doc-collections/insp-gen/2009. The NRC is addressing the survey responses to maintain areas identified as strengths and to improve areas identified as challenges. The staff is developing office and agency action plans and conducting agencywide focus groups to gain further insight into survey findings in order to pursue continuous improvement in both safety culture and organizational effectiveness.

10.5 Managing the Safety and Security Interface

Safety and security have always been the primary pillars of the NRC's regulatory programs.   In today's environment, with a greater emphasis on security-related matters, safety and security activities have become closely intertwined, and it is critical that consideration of these activities be integrated so as not to diminish or adversely impact either safety or security.   While many safety and security activities complement each other, there is the potential for security measures to adversely affect plant safety, and for safety activities to adversely affect security. Recognizing this potential for adverse impact, the NRC has increased its attention to the interfaces between these two areas.

The NRC's mission statement and strategic goals establish a firm foundation for our regulatory framework that stresses the importance of maintaining both safety and security.   The NRC is implementing a number of efforts in the areas of rulemaking, licensing and inspection to recognize, establish and improve this interface.   The NRC has been working multilaterally with the IAEA and bilaterally with our international counterparts to promote this concept.   Since the fourth U.S. National Report was issued, the NRC promulgated 10 CFR 73.58, "Safety/Security Interface Requirements for Nuclear Power Reactors," that requires licensees to assess and manage changes to safety and security activities so as to prevent or mitigate potential adverse affects that could negatively impact plant safety or security.   In addition, as part of the reactor security rulemaking effort, the NRC developed guidance on safety and security interfaces at nuclear power plants, RG 5.74, "Managing the Safety/Security Interface."

The section of this report on major regulatory accomplishments discusses the power reactor security rulemaking in more detail.

In 2000, NRC revised the Reactor Oversight Process to establish a risk-informed baseline inspection program and to set documented risk-informed thresholds for licensee safety and security performance, above which increased NRC oversight would be warranted.   This initiative affirmed the NRC's commitment to better integrate security into the oversight process, by enhancing the safety and security interface as part of the NRC's approach to assess licensee performance.

Satisfactory licensee performance in the Reactor Oversight Process cornerstones provides reasonable assurance of safe and secure facility operation and that the NRC's safety and security missions are being accomplished.   Like the other cornerstones, the security cornerstone contains inspection procedures and performance indicators to ensure that its objectives are being met.   NRC addresses the safety and security interface issues in evaluating their implications among the cornerstones and in the cross-cutting areas of human performance, safety conscious work environment, and problem identification and resolution.   Therefore, safety and security are integrated into the NRC's regulatory framework and evaluated by the NRC staff using a common process.   To ensure licensees are complying with the regulations, the NRC has incorporated the evaluation of the licensee's interfaces with nuclear security into its inspection procedures.

The section of this report on nuclear programs and section 6.3.2 of this report discuss the Reactor Oversight Process in more detail.

Another example of where NRC is promoting strong linkages between safety and security is in the area of organizational culture.   In 2008, the NRC began to expand its policy on safety culture to address the unique aspects of security and to make it applicable to all licensees and certificate holders.   This effort is ongoing and has included interactions and a public meeting with a wide range of stakeholders, including nuclear power plant licensees.

Most participants in the public meeting supported a joint policy statement that addressed safety culture and security culture rather than separate policy statements.   Stakeholders generally believed that the policy statement should recognize that security culture is one of several integrated parts of a licensee's overall safety culture.   In other words, it was recognized that there is no real distinction between cultures, for example, there is not a standalone radiation safety culture, a nuclear criticality safety culture, a fire safety culture, or an environmental protection culture.   Each of these programs is focused on safety for a particular discipline; the licensee safety culture is made up of all the disciplines in an integrated manner.

The resulting safety culture policy statement was submitted to the Commission in SECY 09 0075, "Safety Culture Policy Statement," dated May 18, 2009.   In October 2009, the Commission directed in SRM-SECY-09-0075, "Staff Requirements – SECY-09-0075 – Safety Culture Policy Statement," that the staff publish the policy statement in the Federal Register for public comment. This action will continue to engage a broad range of stakeholders and will seek opportunities to harmonize terminology with existing standards and references.

The section of this report on major regulatory accomplishments and section 10.4 discuss the NRC safety culture in more detail.

ARTICLE 11.   FINANCIAL AND HUMAN RESOURCES

1.   **Each Contracting Party shall take the appropriate steps to ensure that adequate financial resources are available to support the safety of each nuclear installation throughout its life.**

2.   **Each Contracting Party shall take the appropriate steps to ensure that sufficient numbers of qualified staff with appropriate education, training, and retraining are available for all safety-related activities in or for each nuclear installation, throughout its life.**

This section explains the requirements about financial resources that licensees must have to support the nuclear installation throughout its life, and the regulatory requirements for qualifying, training, and retraining personnel.

11.1   Financial Resources

Adequate funds for the safe construction, operation, and decommissioning of nuclear installation are necessary for the protection of public health and safety.   Although there does not appear to be a consistent relationship between a licensee's finances and operational safety, some evidence suggests that financial pressures have limited the resources devoted to corrective actions, plant improvements, and other safety-related expenditures.   Furthermore, because a power reactor must operate to supply the revenues for eventual plant decommissioning, any shutdown of a plant before its owner has accumulated sufficient funds for decommissioning could potentially hinder the safe decommissioning of that plant.

Additionally, many States in the U.S. have undertaken economic deregulation of nuclear power plants.   Traditionally, nuclear power plant owners in many States have been large, vertically integrated companies with substantial assets in generation, transmission, and distribution.   In exchange for having exclusive franchises to supply electric power in defined geographical areas, nuclear plant owners have had the rates they charge to their customers regulated by State government.   This system of rate-based regulation has ensured a source of funds for construction, operation, and decommissioning of nuclear power plants.   Nonetheless, this model of rate-based regulation has been changing and the NRC has adjusted its processes in response.

The NRC distinguishes among financial qualifications for construction, operation, and decommissioning of nuclear power plants, and has separate regulations and programs that apply to each.   The NRC also implements programs to ensure that the public has financial protection for bodily injury and property damage losses in the event of an accident.   Finally, the agency has implemented requirements to ensure that licensees have insurance to help pay onsite recovery costs resulting from accidents and to supply funds for post-accident restart or decommissioning.

11.1.1   Financial Qualifications Program for Construction and Operations

This section explains the financial qualifications program for construction and operations and describes NRC reviews for construction permits, operating licenses, combined licenses, post-operating non-transferred licenses, and license transfers.

Section 182.a of the Atomic Energy Act provides that "each application for a license ... shall specifically state such information as the Commission, by rule or regulation, may determine to be necessary to decide such of the technical and financial qualifications of the applicant ... as the Commission may deem appropriate for the license." To implement this provision, the NRC has developed the regulations and guidance discussed below.

11.1.1.1 Construction Permit Reviews

As required by 10 CFR 50.33(f)(1), applicants for construction permits must submit information that "demonstrates that the applicant possesses or has reasonable assurance of obtaining the funds necessary to cover estimated construction costs and related fuel cycle costs." Appendix C, "A Guide for the Financial Data and Related Information Required to Establish Financial Qualifications for Facility Construction Permits," to 10 CFR Part 50 gives more specific directions for evaluating the financial qualifications of applicants.

11.1.1.2 Operating License Reviews

An "electric utility" as defined in 10 CFR 50.2, "Definitions", is "any entity that generates or distributes electricity and which recovers the cost of this electricity, either directly or indirectly, through rates established by the entity itself or by a separate regulatory authority." Electric utilities are exempt under 10 CFR 50.33(f) from reviews of financial qualifications of applications for operating licenses. The reason for this exemption is that cost-of-service rate regulation, as it has existed in the United States, has ensured that ratepayers provide a source of funds for the safe operation of nuclear power plants. Applicants for operating licenses that are not electric utilities are required under 10 CFR 50.33(f)(2) to submit information that demonstrates that they possess or have reasonable assurance of obtaining the necessary funds to cover estimated operating costs. Nonelectric-utility applicants for operating licenses are also required to submit estimates for the total annual operating costs for each of the first 5 years of operation of their facilities and must state the sources of funds to cover operating costs.

11.1.1.3 Combined License Application Reviews

As authorized in 10 CFR Part 52, applicants may apply for a combined construction permit and operating license. Under 10 CFR 52.77, "Contents of Applications; Technical Information," such applications must contain all of the information required under 10 CFR 50.33, "Contents of Applications; General Information," including information about financial qualifications. The NRC uses the procedures described above to review future combined license applications.

11.1.1.4 Postoperating License Nontransfer Reviews

The NRC does not systematically review the financial qualifications of power reactor licensees once it has issued an operating license, other than for license transfers as described below. However, as provided in 10 CFR 50.33(f)(4), the NRC can seek additional information on licensees' financial resources if the agency considers such information appropriate.

11.1.1.5 Reviews of License Transfers

The NRC regulations in 10 CFR 50.80, "Transfer of Licenses," require agency review and approval of transfers of operating licenses, including licenses for nuclear power plants that are owned or operated by electric utilities. The NRC performs these reviews to determine whether a proposed transferee or new owner is technically and financially qualified to hold the license.

NUREG-1577, Revision 1, "Standard Review Plan on Power Reactor Licensee Financial Qualifications and Decommissioning Funding Assurance," dated February 1999, describes the agency's overall review process of applicant and licensees' financial qualifications for nuclear power plant construction and operation.

11.1.2 Financial Qualifications Program for Decommissioning

Among other sections of the Atomic Energy Act, Section 182.a establishes the basis for the NRC's regulations and guidance on decommissioning funding assurance.   In addition, 10 CFR 50.75, "Reporting and Recordkeeping for Decommissioning Planning," gives the requirements for licensee recordkeeping and reporting of nuclear decommissioning funds to the NRC.

11.1.3 Financial Protection Program for Liability Claims Arising from Accidents

The Price-Anderson Act of 1957, which became Section 170 of the Atomic Energy Act, governs the U.S. financial protection program.   Along with related definitions in Section 11, Section 170 supplies the financial and legal frameworks to compensate those who suffer bodily injury or property damage as a result of accidents at nuclear facilities covered by the law.   The NRC regulations implementing the provisions of Section 170 for NRC licensees are codified in 10 CFR Part 140, "Financial Protection Requirements and Indemnity Agreements."

The Price-Anderson Act was enacted to (1) remove the deterrent to private-sector participation in atomic energy presented by the threat of potentially enormous liability claims in the event of a catastrophic nuclear accident and (2) ensure that adequate funds are available to the public to satisfy liability claims if such an accident were to occur.

The Price-Anderson Act was revised most recently in 2005, when Congress renewed the Commission's authority to cover new facilities until 2020.   Under the current law, power reactors over 100 megawatts electric must contribute to a funding pool that replaces the U.S. Government as the second provider of funds if the first layer of financial protection (liability insurance, now $375 million) is exhausted.

After an accident, reactor operators must pay into a "retrospective premium pool" in maximum annual installments not to exceed $15 million, up to a total of $111.9 million each.   But payment is called for only if the accident exhausts the first layer of financial protection, and only if and to the extent that, additional funds are needed to pay the damages.   With 104 reactors currently participating in the system, the total financial protection available under the Price-Anderson Act for any one accident is approximately $12 billion ($375 million primary coverage plus ($111.9 million per reactor times 104 reactors)) which is also the limit on liability.   As reactors leave the retrospective premium system as a result of permanent closure or join as the result of construction of new reactors, this coverage limit may fall or rise.   A change in the limit may also occur when the $111.9 million contribution is adjusted for inflation, as must be done every 5 years.   In any event, Congress will address any damages exceeding the total sum that reactors must contribute to the pool and will decide upon the next steps needed for compensation.

The public benefits significantly from another feature of the Price-Anderson Act.   Claimants need only prove that the accident caused their injury to receive compensation for damages from any accident with significant offsite releases of radiation (i.e., an "extraordinary nuclear occurrence"). Neither proof of fault nor proof of what caused the accident is necessary.

Claims for more than 150 alleged incidents involving nuclear material have been filed under various liability policies since the inception of the Price-Anderson Act in 1957. The insured losses and expenses paid so far total more than $125 million. Most payments arose out of the accident at Three Mile Island Unit 2.

11.1.4 Insurance Program for Onsite Property Damages Arising from Accidents

Among other sections of the Atomic Energy Act, Section 182.a gives the basis for the NRC's onsite property damage insurance requirements for operating nuclear power reactors contained in 10 CFR 50.54(w).

The U.S. nuclear industry has not experienced an accident involving radioactive release since the Three Mile Island Unit 2 event in 1979.

11.2 Regulatory Requirements for Qualifying, Training, and Retraining Personnel

This section explains the regulatory requirements for qualifying, training, and retraining personnel. It discusses the governing documents, the process for implementing requirements, and experience. It also discusses INPO accreditation activities.

11.2.1 Governing Documents and Process

The NRC regulates the training requirements for licensed operators and licensed senior operators under 10 CFR Part 55, "Operators' Licenses," which allows facility licensees to have operator requalification program content that is derived using a systems approach to training (SAT), as defined in 10 CFR 55.4, "Definitions," or that meets the requirements outlined in 10 CFR 55.59(c). Subpart D, "Applications," of 10 CFR Part 55 requires that operator license applications must contain information about an individual's training and experience, unless the facility licensee certifies that the applicant has successfully completed a Commission-approved training program that is SAT-based and uses an acceptable simulation facility.

The operator licensing process at power reactors includes a generic fundamentals examination covering the theoretical knowledge that is required to operate a nuclear power plant. License applicants must pass the generic fundamentals examination before they can take a site-specific examination. The site-specific examination consists of a written examination and an operating test that includes a plant walkthrough and a dynamic performance demonstration on a simulation facility.

The NRC staff has transferred most of the responsibility for developing site-specific licensing examinations to facility licensees. In 1999, the NRC amended 10 CFR Part 55 to allow nuclear power reactor licensees to prepare the written examinations and operating tests that the agency uses to evaluate the competence of applicants for operators' licenses at those facilities. Licensees that elect to prepare their own examinations are required to establish procedures to control examination security and integrity. They prepare and submit proposed examinations and operating tests to the NRC according to the guidance in NUREG-1021, Revision 9, Supplement 1, "Operator Licensing Examination Standards for Power Reactors," dated October 2007. The NRC reviews the facility-prepared examinations, prepares examinations for facility licensees upon request, administers all operating tests, makes the final licensing decisions, and issues the licenses.

As required by 10 CFR 50.120, "Training and Qualification of Nuclear Power Plant Personnel," licensees must establish, implement, and maintain training programs using a SAT approach for eight categories of non-licensed workers at nuclear power plants and for the shift supervisor, who is licensed in accordance with 10 CFR Part 55. These provisions complement the requirements for training based on a systems approach for the requalification of licensed operators and licensed senior operators. RG 1.8, Revision 3, "Qualification and Training of Personnel for Nuclear Power Plants," dated May 2000, contains guidance to implement the regulations.

The NRC continues to endorse the training accreditation process managed by INPO. The staff recognizes that training programs developed in accordance with INPO guidelines and accredited by the National Nuclear Accrediting Board are SAT based; therefore, accredited programs are considered to be consistent with the regulations in 10 CFR Part 55 and 10 CFR Part 50.120. The NRC also recognizes that INPO-managed accreditation and associated training evaluation activities are an acceptable means of self-improvement in training. Such recognition encourages industry initiative and reduces NRC evaluation and inspection activities.

In accordance with its memorandum of agreement with INPO, the NRC monitors INPO accreditation activities as part of its continuing assessment of the effectiveness of the industry's training programs. Specifically, the NRC staff observes selected accreditation team visits and NRC managers periodically observe National Nuclear Accrediting Board meetings. These observations are intended to monitor the implementation of programmatic aspects of the accreditation process, but they also give an opportunity to assess the selected performance areas of facility licensees.

If the National Nuclear Accrediting Board has concerns about the performance of an accredited training program, it will place the program on probation. This does not necessarily place a training program in non-compliance with either 10 CFR Part 55 or 10 CFR 50.120 because training programs are accredited to a standard of excellence rather than to a minimum level of regulatory compliance. However, the NRC does review the circumstances leading to the probation to ensure safe operations and continued compliance with the regulations.

The National Nuclear Accrediting Board may also withdraw accreditation in response to major deficiencies in a licensee's accredited training program. If accreditation is withdrawn, the NRC would ask that the licensee report the circumstances of the withdrawal for the staff to determine the significance of the issues related to the withdrawal. If the NRC determines that compliance with the regulations is not affected, it may not be necessary to take any further action. If the withdrawal is linked to a breakdown in the training process or a safety-significant issue, the NRC will conduct an immediate inspection focused on the process problem or safety issues. If appropriate, the agency would take further action, such as issuing confirmatory action letters or orders.

The NRC monitors industry performance in implementing the training requirements of 10 CFR Part 50 and 10 CFR Part 55 by (1) reviewing licensee event reports and inspection reports for training issues, (2) observing the accreditation process, and (3) reviewing the results of operator licensing activities. Guidance for periodically inspecting the licensed operator requalification training program at every facility is given in IP 71111.11, "Licensed Operator Requalification Program," dated January 5, 2006. When appropriate for cause, the NRC will also use IP 41500, "Training and Qualification Effectiveness," dated June 13, 1995, which references the guidance in NUREG-1220, Revision 1, "Training Review Criteria and Procedures," dated January 1993, to verify compliance with SAT requirements.

11.2.2 Experience

The NRC reviewed training issues contained in licensee event reports and inspection reports during 2009 using data from the Human Factors Information System, which is described in Article 12.   The review revealed that the proportion of human performance issues attributable to training for U.S. nuclear power plants in 2008 was 4 percent.   As noted in the 2007 version of this report, this figure decreased from 8 percent in 1999 to 4 percent in 2005.   The training-related issues identified by the review concentrated in two subcategories: (1) training less than adequate and (2) individual knowledge less than adequate.   The NRC annually assesses the effectiveness of training in the nuclear industry and prepares a report of its findings; the reports for 1999 through 2007 appear on the NRC's public Web site.

Although the NRC identified some limited specific weaknesses in training programs, all indicators suggest that the industry is successfully implementing training programs in accordance with the regulations.   The NRC will continue to monitor selected performance areas, emphasizing the identification and resolution of training process problems.

ARTICLE 12.   HUMAN FACTORS

Each Contracting Party shall take the appropriate steps to ensure that the capabilities and limitations of human performance are taken into account throughout the life of a nuclear installation.

This section explains the NRC program on human performance.   This program has seven major areas: (1) human factors engineering issues, (2) emergency operating procedures and plant procedures, (3) working hours and staffing, (4) fitness for duty, (5) Human Factors Information System, (6) support to event investigations and for-cause inspections, and (7) training.

12.1   Goals and Mission of the Program

The NRC has a comprehensive program for ensuring that human performance is properly addressed in a risk-informed regulatory framework for maintaining reactor safety.   The NRC developed the program based on reviewing risk information and activities in the domestic and international nuclear industry.

12.2   Program Elements

The Reactor Oversight Process (discussed in Article 6) focuses on cornerstones of safety that are assessed through a combination of performance indicators and risk-informed inspections that focus on risk-significant activities and systems related to the cornerstones.   The three elements that cut across the cornerstones are human performance, a safety-conscious work environment, and corrective actions.   The Human Performance Program has contributed directly to the development of a supplemental IP related to the human performance cross-cutting element.   The Human Performance Program is also engaged in the other two elements, as a safety-conscious work environment and many of the actions involved in corrective action programs result from human performance problems.

The Human Performance Program also supports the risk-informed and performance-based plan by generating, collecting, and evaluating data on human performance for use in human reliability analysis models.   The staff evaluates information to gain insights supporting risk-informed regulation and to find human performance data for human reliability analysis.   The NRC is developing the Human Event Repository and Analysis system to analyze and collect human performance information from commercial nuclear power plants and other related technologies to support regulatory applications in human reliability analysis and human factors.   The system aims to supply empirical evidence to justify or improve human error probabilities in the PRA. The Human Event Repository and Analysis system stores human performance information obtained from event analysis, using the information collection methods and process documented in NUREG/CR-6903, Volumes 1 and 2, "Human Event Repository and Analysis (HERA) System," dated July 2006 and November 2007, respectively.

The Human Performance Program monitors technological developments and emerging issues to help prepare the NRC for the future.   Two ongoing activities include developing regulatory guidance for reviewing designs of control stations and processing requests related to deregulation.   Because licensees are replacing aging analog controls and displays with digital components, the NRC must be prepared to review safety issues for human-system interfaces resulting from such new designs and technologies.   The NRC has been processing many industry requests to transfer operating licenses, which may involve changes in organizational structure affecting human performance.

12.3 Significant Regulatory Activities

The NRC performs significant regulatory activities in the following seven areas to address human performance:

- human factors engineering issues
- emergency operating procedures and plant procedures
- shift staffing
- fitness for duty
- Human Factors Information System
- support to event investigations and for-cause inspections
- training

The following sections cover the first six activities; Article 11 describes training.

12.3.1 Human Factors Engineering Issues

This section discusses human factors activities related to engineering issues.

Governing Documents and Process. The NRC evaluates the human factors engineering design of the main control room and control centers outside of the main control room using NUREG-0800, Chapter 18, Revision 2, "Human Factors Engineering," dated March 2007, NUREG-0700, Revision 2, "Human System Interface Design Review Guideline," dated May 2002, and NUREG-0711, Revision 2, "Human Factors Engineering Program Review Model," dated February 2004.   These documents provide guidance for the review of human-system interface issues in connection with the design certification of nuclear installations and the NRC's inspection program.   The NRC also uses NUREG-1764, Revision 1, "Guidance for the Review of Changes to Human Actions," dated September 2007, to review license amendment requests that credit the use of manual actions.   Moreover, Information Notice (IN) 97-78, "Crediting of Operator Actions in Place of Automatic Actions and Modifications of Operator Actions, Including Response Times," dated October 23, 1997, identifies references that the NRC uses to review the completion times of operator manual actions and how the actions will be reflected in the licensee's emergency procedures and operator training.   In October 2007, the staff published NUREG-1852 "Demonstrating the Feasibility and Reliability of Operator Manual Actions in Response to Fire," for use in evaluating exemptions from fire protection requirements that assume credit for timely manual actions.

In an effort to make some of the current human factors guidance simpler, clearer, and more relevant to the digital environment, the staff published interim staff guidance entitled, "Digital Instrumentation and Controls DI&C-ISG-05 Task Working Group #5 Highly-Integrated Control Rooms—Human Factors Issues (HICR—HF) ISG" Revision 1, dated November 3, 2008, about computer-based procedures, minimum inventory (of controls and displays to support plant shutdown), and crediting manual operator actions in diversity and defense-in-depth analyses. The staff intends to incorporate this interim guidance into permanent regulatory format (such as the standard review plans, NUREGs, RGs, or industry standards) over the next few years.

Experience. The NRC reviews licensees' requests that involve aspects of human factors engineering. Examples include crediting operator manual actions in amendments to plant technical specifications, transferring facility operating licenses, and increasing the reactor's authorized power level (i.e., power uprates). Recent license amendment requests from Oconee Units 1, 2, and 3 and Edwin Hatch Units 1 and 2 are examples of NRC reviews involving new or modified operator manual actions. The amendment from Oconee proposed changes to manual actions as a result of a digital upgrade of the reactor protection system and engineered safety features actuation system. The amendment request from Hatch involved new operator manual actions to support an alternate source term.

The NRC has also evaluated some requests to transfer facility operating licenses, which affected management and organization, staffing, and technical qualifications. The NRC used NUREG-0800, Chapter 13, as the principal guidance for these reviews.

The NRC also reviews and approves requests for power uprates from currently licensed plants. For such requests, the NRC examines the effect of the power uprate on plant procedures, controls, displays, and alarms, and required operator actions using Section 2.11.1 or Review Standard (RS-001), "Review Standard for Extended Power Uprates," dated December 2003. (RS-001 is available on the NRC's public Web site along with additional general information on power uprates.) The agency recently reviewed and approved power plant uprates for Comanche Peak Units 1 and 2, Millstone Unit 3, and Calvert Cliffs.

12.3.2 Emergency Operating Procedures and Plant Procedures

Licensees must have programs to develop, implement, and maintain emergency operating and plant procedures. Article 16 discusses emergency preparedness; the discussion here is limited to the human factors aspect of emergency operating procedures.

Governing Documents and Process. On December 17, 1982, the NRC issued GL 82-33, "Requirements for Emergency Response Capability," which transmitted Supplement 1 to NUREG-0737, "Requirements for Emergency Response Capability," requiring each licensee to submit a set of documents for developing emergency operating procedures.

Experience. No significant examples of emergency operating and plant procedures have been identified since 2007.

12.3.3 Shift Staffing

Governing Documents and Process. In 10 CFR 50.54(m), the NRC specifies the minimum number of licensed operators and senior operators required for nuclear power reactor facilities. Appendix R, "Fire Protection Program for Nuclear Power Facilities Operating Prior to January 1, 1979," and Appendix E, "Emergency Planning and Preparedness for Production and Utilization Facilities," to 10 CFR Part 50 contain the NRC staffing requirements for fire brigades and emergency response personnel.

In September 2002, the NRC began work on a process to evaluate exemption requests from the requirements in 10 CFR 50.54(m) resulting from the changing demands and new technologies presented by advanced reactor control room designs and significant light-water reactor control room upgrades. In July 2005, the NRC published NUREG-1791, "Guidance for Assessing Exemption Requests from the Nuclear Power Plant Licensed Operator Staffing Requirements Specified in 10 CFR 50.54(m)." The purpose of reviewing the exemption requests is to ensure public health and safety by verifying that the applicant's staffing plan and supporting analyses sufficiently justify the requested exemption. NUREG/CR-6838, "Technical Basis for Regulatory Guidance for Assessing Exemption Requests from the Nuclear Power Plant Licensed Operator Staffing Requirements Specified in 10 CFR 50.54(m)," dated February 2004, explains the justification for the recommended process.

Experience. No significant examples of shift staffing issues were identified for 2007–2009.

12.3.4 Fitness for Duty

This section discusses the NRC's requirements pertaining to the fitness for duty of nuclear power plant workers, including requirements regarding the control of work hours and management of worker fatigue.

Governing Documents and Process. As required by 10 CFR Part 26, "Fitness for Duty Programs," each licensee authorized to operate or construct a nuclear power reactor must implement a fitness for duty program for all personnel having unescorted access to the protected area of its plant. For performance objectives, 10 CFR Part 26 requires that licensees establish programs that (1) give reasonable assurance that nuclear power plant personnel perform their tasks in a reliable and trustworthy manner and are not under the influence of any substance, legal or illegal, or mentally or physically impaired from any cause, (2) provide reasonable measures for the early detection of persons who are not fit to perform activities, and (3) have a goal of achieving a drug-free workplace and a workplace free of the effects of such substances.

The NRC issues annual reports on statistical data and lessons learned by licensees from their fitness for duty program performance reports. The most recent of these is IN 2008-16, "Summary of Fitness-for-Duty Program Performance Reports for Calendar Year 2007," dated September 2, 2008. A project to automate the reporting and trending of performance data using a Web-based approach is ongoing. In addition, the NRC has established an email address for licensees and individuals to submit fitness for duty questions, as well as a Web site where performance reports and the answers to frequently asked questions are publicly available.

For worker fatigue, on March 31, 2008, the NRC published a rule that included new regulations in 10 CFR Part 26, Subpart I, "Managing Fatigue." The NRC required licensees to implement the requirements in the rule by October 1, 2009, giving them an 18-month period to hire and train individuals as needed to ensure proper implementation of the work hour control requirements. Subpart I strengthens the effectiveness of fitness for duty programs by ensuring that worker fatigue does not adversely affect public health and safety. It also establishes enforceable requirements for the management of worker fatigue. In addition to the rulemaking and its associated analyses, the NRC issued RG 5.73, "Fatigue Management for Nuclear Power Plant Personnel," in March 2009 to provide guidance on how to implement the rule.

Experience. Licensees have successfully implemented the fitness for duty requirements, as shown by the small number of violations that have occurred to date. However, several issues were identified that needed further staff clarifications and actions. For example, on September 24, 2009, the NRC issued EGM-09-008, "Dispositioning of Violation of NRC Requirements for Work Hours Control Before and Immediately After an Emergency Hurricane Declaration," about staffing before and after a hurricane. Under 10 CFR Part 26, Subpart I, licensees need not meet the work hour control requirements during declared emergencies. The EGM effectively extends this provision by allowing licensees to sequester personnel on site during defined periods before and after a hurricane.

12.3.5 Human Factors Information System

Governing Documents and Process. The Human Factors Information System is designed to store, retrieve, sort, and analyze human performance information extracted from NRC inspection and licensee event reports. Initiated in 1990, this automated information management system can generate a variety of specialized reports that are not readily available from other NRC sources. In 2006, the NRC improved this system to better align the coding scheme with the Reactor Oversight Process and to enhance the system's search capabilities. The Human Factors Information System now captures information related to training, procedures and reference documents, fitness for duty, oversight, problem identification and resolution, communications, human-system interface and environment, and work planning and practices.

Experience. The NRC responds to stakeholder and public inquiries and data requests on this system on a regular basis. For example, inspectors use the data generated by this system in preparing inspection activities related to human performance. In addition, the NRC's Office of Nuclear Regulatory Research uses the data to support activities in human performance and human reliability analysis. Other NRC program offices use the data to gain insights about human performance, to monitor the frequency of human performance issues, and to inform several types of reports, such as internal operating experience reports and the NRC's annual report on the effectiveness of training in the nuclear industry (discussed in Section 11.2.2 of this report). The NRC also uses a Web site to disseminate information on human performance issues at individual nuclear power plant sites.

12.3.6 Support to Event Investigations and For-Cause Inspections and Training

Governing Documents and Process. NRC staff members with human factors expertise often participate in special inspections, incident investigation team inspections, augmented team inspections, event investigations, and supplemental inspections. Human factors experts have assessed management effectiveness, procedures, training issues, staffing issues, human-machine interfaces, personnel performance issues, safety-conscious work environment, and safety culture.

For training issues, inspectors use IP 41500. For procedure issues, inspectors use IP 42001, "Emergency Operating Procedures," dated June 28, 1991, and IP 42700, "Plant Procedures," dated November 15, 1995. For baseline inspections under the Reactor Oversight Process, inspectors use IP 71152, "Problem Identification and Resolution," dated February 26, 2010, which is intended to establish confidence that each licensee is detecting and correcting problems in a way that limits the risk to the public and includes a review of the licensee's safety-conscious work environment. A key premise of the Reactor Oversight Process is that weaknesses in problem identification and resolution programs will manifest themselves as performance issues that can be identified during the baseline inspection program or by crossing predetermined indicator thresholds.

For supplemental inspections, IP 95003, "Supplemental Inspection for Repetitive Degraded Cornerstones, Multiple Degraded Cornerstones, Multiple Yellow Inputs or One Red Input," as revised in October 2006, includes requirements for the NRC staff to review the licensee's third-party safety culture assessment and independently assess the licensee's safety culture. Staff members with technical expertise in human factors and safety culture perform the safety culture inspection activities. The NRC first implemented the revised IP 95003 at the Palo Verde Nuclear Generating Station in October 2007. Based on the lessons learned from the 2007 NRC inspection and on input from the industry and the public, the staff updated Inspection Manual Chapter 0305, "Operating Reactor Assessment Program," in 2009.

Experience.   In 2007, NRC staff with human factors expertise participated in an IP 95003 inspection at Palo Verde to assess human performance at the site.   The inspectors determined that some findings related to procedure adherence had strong human performance contributions. The NRC discussed its safety concerns, and how and when these issues were identified with Palo Verde.   Palo Verde made a commitment to take action to improve their performance.

The NRC increased its plant oversight and conducted numerous inspections.   The results of these inspections demonstrated that performance at Palo Verde had improved substantially.   In March 2009, the NRC determined that the commitments previously made by Palo Verde had been completed and decided to reduce its oversight at this site.

The NRC continued to monitor Palo Verde to verify that the facility is operating safely and that the licensee's performance improvements are being sustained by focusing on the effectiveness of site's programs and processes.   The NRC plans to perform additional inspection activities in selected areas over a 2-year period to monitor Palo Verde improvement initiatives and to look for any indications of potential decline in safety performance at the site.   The first of these inspections was performed in January 2010 to assess the effectiveness of the licensee's corrective actions in addressing the human performance issues identified during the IP 95003 inspection.   The results of this inspection can be found in the Palo Verde Nuclear Generating Station integrated report, dated May 5, 2010.   The NRC staff will perform another inspection in January 2011.

ARTICLE 13.   QUALITY ASSURANCE

Each Contracting Party shall take the appropriate steps to ensure that quality assurance programmes are established and implemented with a view to providing confidence that specified requirements for all activities important to nuclear safety are satisfied throughout the life of a nuclear installation.

This section describes quality assurance requirements and guidance for design and construction, operational activities, and staff licensing reviews.   It also describes quality assurance programs and regulatory guidance.

13.1   Background

Nuclear power facilities must be designed, constructed, and operated in a manner that ensures: (1) the prevention of accidents that could cause undue risk to public health and safety and (2) the mitigation of adverse consequences of such accidents if they should occur.   A primary means to achieve these objectives is to establish and effectively implement a nuclear quality assurance program.   Although a licensee may delegate aspects of the establishment or execution of the quality assurance program to others, the licensee remains ultimately responsible for the program's overall effectiveness.   Licensees carry out a variety of self-assessments to validate the effectiveness of their quality assurance program implementation.   The NRC reviews descriptions of quality assurance programs and performs onsite inspections to verify aspects of the program implementation.

13.2   Regulatory Policy and Requirements

The NRC describes requirements for a license to design, construct, and operate commercial nuclear power plants in both 10 CFR Part 50 and 10 CFR Part 52.   Specifically, 10 CFR Part 50 describes the requirements for a construction permit and a separate operating license and 10 CFR Part 52 includes the requirements for a single combined license, which allows for both construction and operation of a nuclear power plant.

For either type of license, an applicant must describe its quality assurance program for all activities affecting the safety-related functions of SSCs that prevent or mitigate the consequences of postulated accidents that could cause undue risk to public health and safety. High-level criteria for determining which plant SSCs are safety-related appear in 10 CFR 50.2. Based upon these criteria, licensees' engineering organizations develop plant-specific listings of safety-related SSCs.

Under 10 CFR Part 50 licensing process, each applicant for a construction permit must describe its quality assurance program in its preliminary safety analysis report in accordance with 10 CFR 50.34(a)(7).   This program should apply to the design, fabrication, construction, and testing of SSCs.   In accordance with 10 CFR 50.34(b)(6)(ii), each applicant for an operating license under 10 CFR Part 50 must describe the managerial and administrative controls that will be implemented during the operation of the nuclear power plant.   The applicant must also describe how it will satisfy the applicable requirements of 10 CFR Part 50, Appendix B, "Quality Assurance Criteria for Nuclear Power Plants and Fuel Reprocessing Plants."

Each applicant for a combined license under 10 CFR Part 52 must describe its quality assurance program in a safety analysis report and give a description of the managerial and administrative controls that will be implemented during the operation of the nuclear power plant. Like a 10 CFR Part 50 applicant, an applicant under 10 CFR Part 52 must also describe how it will satisfy the applicable requirements of 10 CFR Part 50, Appendix B.

13.2.1 Appendix A to 10 CFR Part 50

Appendix A, "General Design Criteria for Nuclear Power Plants," to 10 CFR Part 50 details the general requirements for establishing quality assurance controls. General Design Criterion 1, "Quality Standards and Records," contains requirements that apply to the quality assurance of items important to safety. The scope of items that are "important to safety" includes a subset of plant equipment classified as safety-related. Appendix B to 10 CFR Part 50 (discussed in Section 13.2.2 of this report) contains quality assurance program requirements for safety-related SSCs. Other regulatory guidance discusses quality assurance program controls that are appropriate for some types of nonsafety-related equipment.

13.2.2 Appendix B to 10 CFR Part 50

Appendix B to 10 CFR Part 50 outlines the quality assurance requirements that apply to activities affecting the safety-related functions of SSCs that prevent or mitigate the consequences of postulated accidents. Appendix B defines quality assurance as all planned and systematic actions that are necessary for adequate confidence that SSCs will perform satisfactorily in service. Toward that end, it specifies 18 criteria that the commitments in a licensee's quality assurance program must satisfy. These criteria cover such topics as organizational independence, design control, procurement, document control, test control, corrective action, and audits. Appendix B also stipulates that licensees establish measures to ensure that the documents for procurement of safety-related materials, equipment, and services, whether purchased by the licensee or its contractors or subcontractors, include or reference the applicable regulatory requirements, design bases, and other requirements that are necessary to ensure adequate quality. Consistent with the importance and complexity of the products or services to be provided, licensees (or their designees) are responsible for periodically verifying that suppliers' quality assurance programs comply, as appropriate, with the applicable criteria in Appendix B and that they are effectively implemented. Additionally, as outlined in 10 CFR 21.41, "Inspections," the NRC staff performs inspections at vendors who supply basic components to the nuclear industry.

Because the requirements of Appendix B are written at a conceptual level, the NRC and the industry needed to develop consensus standards that include acceptable ways to conform to these requirements. The NRC then issued companion RGs, which endorsed (with conditions, if warranted) quality assurance codes and standards.

13.2.3 Approaches for Adopting More Widely Accepted International Quality Standards

The NRC has reviewed options for adopting more widely accepted international quality standards, such as International Organization for Standardization Standard 9001, 2000 edition, by considering how international standards compare with the existing framework in 10 CFR Part 50, Appendix B. On the basis of this review, the NRC concluded that supplemental quality requirements would be needed when implementing Standard 9001 within the existing regulatory framework. The NRC participates in both national and international efforts associated with quality assurance standard development and it continues to assess how various national and international quality standards comport with NRC regulations in an ongoing effort to seek convergence of standards.

13.3 Quality Assurance Regulatory Guidance

The NRC has developed or endorsed quality assurance guidance for use by the NRC staff, applicants for construction permits or operating licenses, and licensees.   This guidance is applicable to the design, construction, and operational phases of a nuclear power plant.

13.3.1 Guidance for Staff Reviews for Licensing

NUREG-0800, Section 17.5, "Quality Assurance Program Description – Design Certification, Early Site Permit and New License Applicants," dated March 2007, provides guidance to the NRC staff for the review of applications for construction permits, operating licenses, and combined licenses.   The specific review guidance in NUREG-0800 correlates with the 18 criteria of 10 CFR Part 50, Appendix B and integrates a review of licensee commitments to adopt the NRC's quality assurance-related RGs and apply the industry's quality assurance codes and standards.

13.3.2 Guidance for Design and Construction Activities

Licensees may apply consensus standards developed by the American National Standards Institute (ANSI) in its N45.2 series or by ASME in its NQA-1 series to comply with the requirements of 10 CFR Part 50, Appendix B.   The NRC has endorsed ANSI and ASME standards through its RGs.   Through its consensus codes and standards activities, the NRC continues to participate with ASME NQA-1 committees to revise the latest edition of the NQA-1 standard.   As part of this effort, the NRC staff is planning to issue a revision to RG 1.28, "Quality Assurance Program Requirements (Design and Construction)," dated August 1985, to endorse NQA-1-2008 and the 2009-10 addenda.

13.3.3 Guidance for Operational Activities

The NRC has conditionally endorsed the consensus standard ANSI N18.7-1976, "Administrative Controls and Quality Assurance for the Operational Phase of Nuclear Power Plants" through RG 1.33, Revision 2, "Quality Assurance Program Requirements (Operations)," dated February 1978, as complying with the requirements of 10 CFR Part 50, Appendix B.

13.4 Quality Assurance Programs

The NRC inspects quality assurance programs under the Reactor Oversight Process for operating reactors and under the Construction Inspection Program (see Article 18 of this report) for new reactors.   The NRC also conducts augmented inspection activities as needed.

The baseline inspection program of the Reactor Oversight Process includes one primary procedure related to quality assurance issues, IP 71152.   Inspectors use this procedure to assess the effectiveness of licensees' programs to find and resolve problems through a performance-based review of specific issues.   In particular, inspectors look for cases in which a licensee may have missed generic implications of specific problems and for the risk significance of combinations of problems that individually may not have significance.   They do not inspect other aspects of quality assurance program implementation in the baseline inspection program but may do so through supplemental inspections.

Some equipment in the nuclear facility may be classified as nonsafety-related and yet still be important to safety for some unique reason. In specific cases, the NRC has specified that quality assurance controls are warranted for equipment determined to be more important than commercial-grade equipment. However, the quality assurance controls do not have to meet Appendix B requirements, which apply only to activities affecting safety-related functions. Typically, applying quality assurance controls to this important-to-safety, yet nonsafety-related, equipment is called "augmented quality control."

The Construction Inspection Program provides oversight for future nuclear plants licensed under 10 CFR Part 52, including quality assurance program inspection. The quality assurance inspection program focuses on an applicant or licensee establishing and implementing a quality assurance program in accordance with the requirements of Appendix B to 10 CFR Part 50.

As provided in the Construction Inspection Program, the nuclear plant will transition from the Construction Inspection Program to the Reactor Oversight Process for commercial operation when, in accordance with 10 CFR 52.103(g), the Commission determines that all of the inspections, tests, analyses, and acceptance criteria in the combined license have been met.

13.5 Quality Assurance Audits Performed by Licensees

Appendix B to 10 CFR Part 50 requires licensees to verify the effectiveness of their quality assurance program by performing internal audits of their programs. These audits are performed in accordance with the licensee's procedures by appropriately trained and qualified personnel who do not have direct responsibility for performing the activities being audited. The results of these audits are documented and given to management for review and corrective action.

13.5.1 Audits of Vendors and Suppliers

Appendix B to 10 CFR Part 50 requires licensees who procure material, equipment, or services from contractors or subcontractors to perform audits to ensure that suppliers implement an effective quality assurance program, consistent with the requirements of Appendix B and the licensee's technical requirements.

Licensees perform these activities by using their own technical and quality assurance staff. Industry initiatives to promote effective and efficient standardization of these audit activities have resulted in licensees sharing their technical resources through joint audits of suppliers.

ARTICLE 14. ASSESSMENT AND VERIFICATION OF SAFETY

Each Contracting Party shall take the appropriate steps to ensure that:

(i)   comprehensive and systematic safety assessments are carried out before the construction and commissioning of a nuclear installation and throughout its life. Such assessments shall be well documented, subsequently updated in the light of operating experience and significant new safety information, and reviewed under the authority of the regulatory body

(ii)   verification by analysis, surveillance, testing, and inspection is carried out to ensure that the physical state and the operation of nuclear installations continue to be in assurance with its design, applicable national safety requirements, and operational limits and conditions

This section explains the governing documents and process for ensuring that systematic safety assessments are carried out during the life of the nuclear installation, including for the period of extended operation.   It focuses on assessments performed to maintain the licensing basis of a nuclear installation.   Finally, this section explains verification of the physical state and operation of the nuclear installation by analysis, surveillance, testing, and inspection.

Other articles in this report (e.g., Articles 6, 10, 13, 18, and 19) also discuss activities to achieve safety at nuclear installations.

14.1   Ensuring Safety Assessments throughout Plant Life

Before a nuclear facility is constructed, commissioned, and licensed, an applicant must perform comprehensive and systematic safety assessments for NRC review and approval.   Article 18 of this report discusses these assessments and reviews.

This section focuses on the assessments that are required throughout the life of a nuclear installation (i.e., assessments required to maintain the licensing basis).   To show conformance with the licensing basis, a licensee must maintain records of the original design bases and any changes.   This section explains how such changes are documented, updated, and reviewed. Renewal of a license depends on a licensee's continuing to meet its current licensing basis; this section explains how the license renewal process accounts for this requirement.

14.1.1   Maintaining the Licensing Basis

The NRC carries out regulatory programs to give reasonable assurance that plants continue to conform to the licensing basis.   Article 6 of this report discusses these programs.

This section explains the governing documents and process used to maintain the licensing basis. The main governing documents are 10 CFR 50.90, "Application for Amendment of License or Construction Permit, or Early Site Permit," 10 CFR 50.59, "Changes, Tests, and Experiments," and 10 CFR 50.71, "Maintenance of Records, Making of Reports."

14.1.1.1 Governing Documents and Process

A licensee is to operate its facility in accordance with the license and as described in its final safety analysis report. To change its license or reactor facility, a licensee must follow the review and approval processes established in the regulations. For license amendments, including changes to technical specifications, the licensee must ask for NRC approval in accordance with 10 CFR 50.90. However, 10 CFR 50.59 contains requirements for the process by which, under certain conditions, licensees may make changes to their facilities and procedures as described in the safety analysis report without prior NRC approval.

10 CFR 50.59. In 10 CFR 50.59 the NRC establishes the conditions under which licensees may make changes to the facility or procedures and conduct tests or experiments without prior NRC approval. Proposed changes, tests, and experiments that satisfy the definitions and one or more of the criteria in the rule must be reviewed and approved by the NRC before implementation. Thus, the rule provides a threshold for regulatory review, not the final determination of safety, for proposed activities. After determining that a proposed activity is safe and effective through appropriate engineering and technical evaluations, the 10 CFR 50.59 process is applied to determine if a license amendment will be required before implementation. The process involves three basic steps: (1) applicability and screening to determine if a 10 CFR 50.59 evaluation is required, (2) an evaluation that applies the eight evaluation criteria of 10 CFR 50.59(c)(2) to determine if a license amendment must be obtained from the NRC, and (3) documentation and reporting to the NRC of activities implemented under 10 CFR 50.59.

A licensee shall obtain a license amendment pursuant to 10 CFR 50.90 before implementing a proposed change, test, or experiment if the change, test, or experiment would do any of the following:

- result in more than a minimal increase in the frequency of occurrence of a previously evaluated accident
- result in more than a minimal increase in the likelihood of occurrence of a malfunction of an SSC important to safety
- result in more than a minimal increase in the consequences of a previously evaluated accident
- result in more than a minimal increase in the consequences of a malfunction of an SSC important to safety
- create a possibility for an accident of a different type than any previously evaluated
- create a possibility for a malfunction of an SSC important to safety with a different result than any previously evaluated
- result in exceeding or altering a design basis limit for a fission product barrier
- result in a departure from a method of evaluation used in establishing the design bases or in the safety analyses

According to 10 CFR 50.90, whenever a holder of a license or construction permit wants to amend the license or permit, it must file an application for an amendment with the Commission, as specified in 10 CFR 50.4, "Written Communications," fully describing the changes desired, and following, as far as applicable, the form prescribed for original applications. The NRC performs and documents a safety evaluation in these instances before it authorizes the change.

10 CFR 50.71.   In 10 CFR 50.71(e), the NRC describes another process for making changes. This regulation requires licensees to update their final safety analysis reports periodically to incorporate the information and analyses that they submitted to the Commission or prepared pursuant to Commission requirements.   Revisions to the updated final safety analysis reports are to include the effects of changes that occur in the vicinity of the plant, changes made in the facility or procedures described in the report, safety evaluations for approved license amendments and for changes made under 10 CFR 50.59, and safety analyses conducted at the request of the Commission to address new safety issues.

14.1.1.2   Regulatory Framework for the Restart of Browns Ferry Unit 1

As an example of the application of the regulatory framework, this section describes the safety assessment and verification for a plant that was restarted after being shut down for some years.

The Browns Ferry site, located near Decatur, AL, has three BWRs (General Electric (GE), BWR-4, Mark-1 containment).   All three units were shut down in 1985 to address management and regulatory issues.   After resolving these issues, TVA successfully restarted Units 2 and 3 in the 1990s, but kept Unit 1 in a defueled layup condition.   In May 2002, TVA decided to initiate a restart effort for Unit 1.   The three Browns Ferry units are similar in design and licensing basis. TVA has implemented programs for Unit 1 that are similar to those used to restart Units 2 and 3, incorporating improvements, lessons learned, and dedicated resources, including personnel with experience restarting Units 2 and 3.   The restart of Unit 1 differed from the restart of Units 2 and 3 in that TVA applied simultaneously for both a license renewal and an extended power uprate for the unit.

The regulatory framework for the restart of Browns Ferry Unit 1 consisted of two major elements: inspection and licensing activities.   The NRC performed inspections in accordance with Inspection Manual Chapter 2509, "Browns Ferry Unit 1 Restart Project Inspection Program," dated September 2003, and conducted the licensing activities consistent with its August 2003 regulatory framework letter discussed below.

TVA has submitted many and varied licensing actions over the years.   The August 2003 regulatory framework letter included a detailed list of generic communications and other licensing actions requiring regulatory review, approval, and follow-up inspection.   To facilitate communication with key stakeholders, the NRC held periodic public meetings at the site and developed a public outreach Web page similar to the Reactor Oversight Process Web page.

As part of its inspection program, the NRC staff reviewed TVA programs and plant activities related to the recovery of Unit 1.   These activities included replacement, renovation, and removal of equipment and a review of plant programs, process, and training of plant personnel. The NRC inspections of structural, electrical, mechanical, and fire protection modifications resulted in satisfactory findings.   Onsite monitoring and review determined that activities involving replacement, renovation, and removal of equipment were satisfactorily carried out so as to maintain adequate nuclear and radiological safety.

The NRC also conducted an operational readiness assessment team inspection in April 2007 to assess management controls, implementation of site programs and personnel readiness to support safe restart and operation of Unit 1.   The inspection focused on the effectiveness of licensee management oversight, safety-significant activities, operator training and experience, corrective action programs, maintenance program, operator response to annunciators and general plant conditions affecting safety, and readiness to support three-unit operations. The inspection concluded that site programs, personnel, and procedures were adequate for restart of Unit 1 and three-unit power operations.

On May 15, 2007, the NRC authorized TVA to restart Browns Ferry Unit 1. The unit was restarted on May 22, 2007, and reached 100-percent power on June 8, 2007. TVA completed post-restart testing.

After extensive reviews, inspections, and resolution of regulatory framework issues, the NRC is now conducting oversight for Browns Ferry in accordance with its Reactor Oversight Process. However, because of the lack of valid historical data specific to this plant for the mitigating systems performance indicators of the Reactor Oversight Process, the NRC will conduct additional Reactor Oversight Process baseline inspections until sufficient plant-specific data become available in the third quarter of calendar year 2010.

14.1.2   License Renewal

This section explains license renewal, including the governing documents, regulatory process, recent experience, and relevant examples.

14.1.2.1   Governing Documents and Process

Background. The Atomic Energy Act and NRC regulations limit commercial power reactor licenses to 40 years but permit such licenses to be renewed. The original 40-year term was selected on the basis of economic and antitrust considerations, not technical limitations.

The NRC has established a license renewal process that can be completed in a reasonable time period and has clear requirements to ensure safe plant operation for up to 20 additional years of plant life. The NRC's current schedule is to complete renewal reviews within 30 months of receipt of the application if a hearing is conducted, and within 22 months if a hearing is not conducted. Currently, five applications are in the hearing process, and two applications are experiencing extended reviews. The decision to seek license renewal rests entirely with nuclear power plant owners and typically is based on the plant's economic situation and whether it can meet NRC requirements.

Research has concluded that aging phenomena are readily manageable and do not pose technical issues that would prevent life extension for nuclear power plants. Studies have also found that facilities deal adequately with many aging effects during the initial license period, and that credit should be given for these existing programs, particularly those under the NRC's Maintenance Rule (10 CFR 50.65, "Requirements for Monitoring the Effectiveness of Maintenance at Nuclear Power Plants"), which helps manage plant aging.

The license renewal process proceeds along two tracks: one for the review of safety issues and another for environmental issues. An applicant must give the NRC an evaluation that addresses the technical aspects of plant aging and describes the ways it will manage those effects. It must also prepare an evaluation of the potential impact on the environment if the plant operates for up to 20 more years. The NRC reviews the application and verifies the safety and environmental issues through on-site audits and inspections. The NRC documents its findings in a safety evaluation report and an environmental impact statement.

Public participation is an important part of the license renewal process. Members of the public have opportunities to comment on the environmental review and question how aging will be managed during the period of extended operation. All information related to the review and approval of a renewal application is publicly available. Significant safety and environmental concerns may also be litigated in an adjudicatory hearing if any party who would be adversely affected asks for a hearing.

10 CFR Part 54. Known as the License Renewal Rule, 10 CFR Part 54 establishes the technical and procedural requirements for renewing operating licenses. License renewal requirements for power reactors are based on two key principles:

(1)   When continued into the extended period of operation, the regulatory process, which assesses and verifies safety, is adequate to ensure that the licensing basis of all currently operating plants provides an acceptable level of safety. The possible exception is detrimental effects of aging on certain SSCs, and possibly a few other issues applying to safety only during the period of extended operation.

(2)   Each plant must maintain its licensing basis throughout the renewal term.

Guidance that applies to license renewal includes RG 1.188, Revision 1, "Standard Format and Content for Applications to Renew Nuclear Power Plant Operating Licenses," dated September 2005, to help applicants apply to renew a license; and NUREG-1800, Revision 1, "Standard Review Plan for Review of License Renewal Applications for Nuclear Power Plants," dated September 2005, which guides the staff in reviewing applications. The standard review plan for license renewal incorporates by reference NUREG-1801, Revision 1, "Generic Aging Lessons Learned (GALL) Report," dated September 2005, which generically documents the basis for determining when existing programs are adequate for license renewal and when they should be augmented. As lessons are learned from the review of renewal applications or generic technical issues are resolved, the NRC issues improved guidance for interim use by applicants until the guidance is incorporated into the next formal update of the documents. The staff is currently preparing a revision to both the standard review plan for license renewal and the Generic Aging Lessons Learned Report. The NRC obtained comments from the public on these documents and plans to issue them for use by December 2010.

10 CFR Part 51. The NRC's environmental protection regulation, 10 CFR Part 51, "Environmental Protection Regulations for Domestic Licensing and Related Regulatory Functions," also applies to license renewal. The agency amended this regulation to facilitate its environmental review process for license renewal. The review requirements for 10 CFR Part 51 are founded on the conclusion that certain environmental issues can be resolved generically and need not be evaluated in each plant-specific application. NUREG-1437, "Generic Environmental Impact Statement for License Renewal of Nuclear Plants," dated May 1996, describes these issues. The NRC performs plant-specific reviews of the environmental impacts of license renewal to determine whether the effects are so great that they should preclude license renewal as an option for energy-planning decisionmakers.

RG 4.2, Supplement 1, "Preparation of Supplemental Environmental Reports for Applications to Renew Nuclear Power Plant Operating Licenses," dated August 1991, provides guidance to applicants preparing environmental reports for license renewal. NUREG-1555, "Standard Review Plans for Environmental Reviews for Nuclear Power Plants, Supplement 1, "Operating License Renewal," dated March 2000, guides the NRC staff's review of the environmental issues associated with a renewal application. The NRC, with public participation, is currently revising its regulations and guidance in this area. The rulemaking proposing changes to Part 51 was issued in the *Federal Register* in 2009. The staff expects to issue the final rulemaking in 2011.

14.1.2.2   Experience

The NRC issued the first renewed licenses for the Calvert Cliffs Nuclear Power Plant and the Oconee Nuclear Station in 2000. As of March 2010, 59 reactors have received renewed licenses. Four of the 59 reactors have completed 40 years of operation and are operating in the extended period. Three more reactors will enter the period of extended operation in the second half of 2010. On the basis of industry statements, the NRC expects that essentially all remaining plants will apply for license renewal.

14.1.2.3 Operating Beyond 60 Years

The provisions of 10 CFR Part 54 do not preclude subsequent license renewals after the initial renewal. The earliest that a licensee can submit a license renewal application is 20 years before the expiration of its current license; therefore, a licensee is eligible to apply for a subsequent license renewal once it enters the initial period of extended operation (the 20-year renewal period beyond its initial 40-year license period). While several industry representatives have informally inquired about the possibility of license renewal beyond 60 years, the Commission has not received any formal letter of intent to pursue such a renewal.

To encourage early and proactive discussion of factors potentially affecting subsequent license renewal decisions, the Commission and DOE jointly sponsored a workshop on U.S. nuclear power plant life extension research and development on February 19 – 21, 2008. Based on the results of the workshop and the staff's long-term research plan, potential additional areas of focus for a subsequent license renewal include aging management of reactor vessel and internal materials, cable insulation, buried piping, submerged structures, and concrete exposed to high temperature and radiation. It is the industry's responsibility to conduct the necessary research to support a request for a second, subsequent license renewal. The NRC's Office of Nuclear Reactor Regulation is also closely coordinating with the NRC's Office of Nuclear Regulatory Research to track industry work in this area, evaluate areas for research, and gather data to help in assessing the effectiveness of licensee's aging management programs.

14.1.3   The United States and Periodic Safety Reviews

The international community, to a large extent, conducts periodic safety reviews (typically carried out every 10 years) to assess the cumulative effects of plant aging, plant modifications, operating experience, technical developments, and siting. The reviews include an assessment of plant design and operation against current safety standards and practices, with the objective of ensuring a high level of safety throughout the plant's operating lifetime.

Some countries use routine comprehensive safety assessment programs that deal with specific safety issues, significant events, and changes in safety standards and practices as they arise.

These programs, if applied with appropriate scope, frequency, depth, and rigor, achieve the same review standards and objectives as a periodic safety review. Some countries also use periodic safety reviews to support the decisionmaking process for long-term operation or license renewal. However, alternate processes, such as the NRC license renewal process, are considered equally adequate and acceptable.

This section explains how the U.S. regulatory approach provides a continuum of assessment and review that ensures public health and safety throughout the period of plant operation. Plant safety is maintained, and aspects are improved, by a combination of the ongoing NRC regulatory process, oversight of the current licensing basis, backfitting, broad-based evaluations, license renewal, and licensee initiatives that go beyond the regulations.

14.1.3.1   The NRC's Robust and Ongoing Regulatory Process and the Current Licensing Basis

Before issuing an operating license, the NRC determines that the design, construction, and proposed operation of the nuclear power plant satisfy the NRC's requirements and reasonably ensure the adequate protection of public health and safety. However, the licensing basis of a plant does not remain fixed for the 40-year term of the operating license. The licensing basis evolves throughout the term of the operating license because of the NRC's continuing regulatory activities and the licensee's activities.

The NRC carries out many regulatory activities that, when considered together, constitute a process providing ongoing assurance that the licensing bases of nuclear power plants provide an acceptable level of safety. This process includes inspections (both periodic regional inspections as well as daily oversight by the resident inspectors), audits, investigations, evaluations of operating experience, regulatory research, and regulatory actions to resolve identified issues. The NRC's activities may result in changes to the licensing basis for nuclear power plants through promulgation of new or revised regulations, acceptance of licensee commitments to modify nuclear power plant designs and procedures, and the issuance of orders or confirmatory action letters. The agency also publishes the results of operating experience analysis, research, or other appropriate analyses through generic communication documents such as bulletins, INs and GLs. Licensee commitments in response to these documents also change the plant's licensing basis. In this way, the NRC's consideration of new information gives ongoing assurance that the licensing basis for the design and operation of all nuclear power plants provides an acceptable level of safety. This process continues for plants that receive a renewed license to operate beyond the original operating license.

In addition to NRC-required changes in the licensing basis, a licensee may also voluntarily seek changes to the current licensing basis for its plant. These changes are subject to the NRC's formal regulatory controls on changes (such as those described in 10 CFR 50.54, "Conditions of Licenses," 10 CFR 50.59, 10 CFR 50.90, and 10 CFR 50.92, "Issuance of Amendment"). These regulatory controls ensure that licensee-initiated changes to the licensing basis are documented and that the licensee obtains NRC review and approval, if necessary, before implementing them. The licensee must report to the NRC any changes or modifications it makes to the licensing basis without prior NRC review at least every 2 years. Region-based NRC inspectors perform a sampling inspection of those changes in accordance with the Reactor Oversight Process to ensure that the licensee has properly characterized the changes or modifications.

14.1.3.2 The Backfitting Process: Timely Imposition of New Requirements

In the late 1970s and early 1980s, the NRC recognized the need for a process to determine when to address generic issues for all plants. The NRC deemed prudent to consider new requirements systematically rather than depending on other regulatory processes to decide on plant upgrades. As a result, the NRC developed the "backfitting" process and established the Committee to Review Generic Requirements to review staff-proposed backfits on licensees.

The Backfitting Rule, 10 CFR 50.109, "Backfitting," promulgated in 1985, applies to both generic and plant-specific backfits for power reactors. The rule defines a "backfit" as any modification of or addition to (1) plant systems, (2) structures, (3) components, (4) design approvals, (5) manufacturing licenses, or (6) procedures or organization required to design, construct, or operate a facility that may result from the imposition of a new or amended rule or regulatory staff position.

In 1988, the NRC amended the Backfitting Rule to state that economic costs will not be considered in cases of ensuring, defining, or redefining adequate protection of public health and safety, or in cases of ensuring compliance with NRC requirements or written licensee commitments. The rule requires a cost-benefit analysis except in the case of backfits that are imposed to bring a licensee back into compliance with its license or to ensure adequate protection of public health and safety or the common defense and security. The NRC must determine through a backfit analysis that the proposed backfit will substantially increase the overall protection of public health and safety or the common defense and security and that the direct and indirect costs for the facility are justified in view of the increased protection.

Compliance and adequate protection backfits are justified differently. The NRC requires a documented evaluation that gives the basis and states the objectives and purpose of the proposed backfit.

Backfitting is expected and is an inherent part of the regulatory process. However, it is permitted only after a formal, systematic review to ensure that changes are properly justified and suitably defined. The requirements of this process are intended to ensure order, discipline, and predictability and to optimize the use of NRC staff and licensee resources.

The controls on generic backfitting include review by the Committee to Review Generic Requirements, a committee of senior managers from various NRC offices. Established in 1981, this committee operates under a charter that specifically identifies the documents to be reviewed and the analyses, justifications, and findings to be supplied. Its objectives include eliminating unnecessary burdens on licensees, reducing radiation exposure to workers while implementing requirements, and optimizing use of NRC and licensee resources to ensure safe operation. Thus, the Committee to Review Generic Requirements charter is a key implementing procedure for generic backfitting, although the primary responsibility for proper backfit considerations belongs to the initiating organization.

14.1.3.3 The NRC's Extensive Experience with Broad-Based Evaluations

In the mid-1970s, the NRC recognized the importance of assessing the adequacy of the design and operation of currently licensed nuclear power plants, understanding the safety significance of deviations from applicable current safety standards that may have been approved after those plants were licensed, and providing the capability to make integrated and balanced decisions about the need for backfit modifications at those plants.

Consequently, in 1977, the NRC initiated the Systematic Evaluation Program (SEP). From a list of approximately 800 potential issues and topics related to nuclear safety, the SEP found that the regulatory requirements for 137 issues had changed sufficiently to warrant evaluation. The staff compared the designs of 10 of the older plants to the licensing criteria delineated in the then recently issued standard review plan.[4] After further review, the staff determined that 27 issues required some corrective action at one or more plants and that resolution of those issues could lead to safety improvements at other operating plants built at about the same time. These 27 issues became known as the "27 SEP lessons learned."

In 1984, NRC staff presented the 27 SEP lessons learned to the Commission as part of a proposal for an Integrated Safety Assessment Program (ISAP). The staff developed this program to review safety issues for a specific plant in an integrated manner instead of continuing the SEP at other older operating reactors. In "Commission Policy Statement on the Systematic Evaluation of Operating Nuclear Power Reactors," dated November 1984, the Commission said that issues relating to the safety of operating nuclear power plants can be more effectively and efficiently implemented in an integrated, plant-specific review. For the first time, the Commission discussed probabilistic safety analysis as a method to obtain consistent and comparable results that could be used to enhance a safety assessment. The SEP process was transformed into the ISAP pilot program.

In May 1985, the NRC initiated the ISAP pilot at two plants, Millstone Unit 1 and Haddam Neck (Connecticut Yankee). The ISAP pilot identified some benefits; however, the Commission deferred extending it beyond the pilot phase until the staff gave an integrated package of options that clarified the relationship between the proposed follow-on program to the ISAP pilot (ISAP II) and the newly proposed individual plant examination process.

The Commission determined that, since ISAP II would be voluntary and the individual plant examination program, through the NRC's GL process, would require a licensee response, the staff should give priority to the individual plant examination program. Many of the same benefits that might have been derived through the proposed ISAP II were derived instead through the individual plant examination process (e.g., probabilistic safety analysis).

In the late 1980s and throughout the 1990s, the NRC continued to strengthen its regulatory infrastructure and ensure the continued safe operation of commercial nuclear power plants through inspection, broad-based assessment, and, where appropriate, establishment of new generic requirements. For example, the Commission determined that licensees should assess the accessibility and adequacy of their design-basis information and determine whether their plants needed a design-basis reconstitution program. The Commission expressed its expectations in "Availability and Adequacy of Design Bases Information at Nuclear Power Plants; Policy Statement" in the *Federal Register* on August 10, 1992. The Commission also expanded the individual plant examination program to consider external events and, recognizing the

---

[4]   Standard review plans help ensure the quality and uniformity of staff reviews and provide a well-defined base from which to evaluate a licensee or applicant submittal. Standard review plans are also intended to make information about regulatory matters widely available, to enhance communication with interested members of the public and the nuclear power industry, and to improve the understanding of the staff review process.

relationship between maintenance, equipment reliability, plant risk, and safety, in 1991 the Commission promulgated the Maintenance Rule codified in 10 CFR 50.65.

The Maintenance Rule requires licensees to monitor the performance or condition of SSCs against licensee-established goals continuously, to give reasonable assurance that these SSCs are capable of fulfilling their intended functions. The NRC verifies the licensee's implementation of the Maintenance Rule through the Reactor Oversight Process, periodic regional inspections, and daily oversight by the resident inspectors.

As late as 1991, some plants had not definitively resolved the 27 SEP lessons learned. As the staff considered a process to renew the operating licenses for the operating nuclear power plants, it assessed the best way to address these 27 issues.

Of the 27 issues, four had been completely resolved for all plants. One other issue was of such low safety significance that it required no additional action. The staff determined that none of the remaining 22 issues required immediate action to protect public health and safety. The staff placed these 22 issues into the established regulatory process for determining the safety significance of generic issues.[5]

14.1.3.4 License Renewal Confirms Safety of Plants

In developing the License Renewal Rule, the Commission concluded that issues material to the renewal of a nuclear power plant operating license are limited to those issues that the Commission determines are uniquely relevant to protecting public health and safety and preserving the common defense and security during the period of extended operation. Other issues would, by definition, be relevant to the safety and security of the public during current plant operation. Given the Commission's ongoing obligation to oversee the safety and security of operating reactors, the existing regulatory process within the present 40-year license term addresses issues related to current plant operation rather than deferring the issues until the time of license renewal. The NRC manages these issues by implementing the Reactor Oversight Process, generic communications, and the generic safety issues program. (Section 6.3.2 of this report describes the NRC Reactor Oversight Process.)

---

[5] A generic issue is a regulatory matter that is not sufficiently addressed by existing regulations, guidance, or programs. Through its systematic assessment of plant operation, the NRC has identified certain issues that seem prevalent among plants. The NRC documents and tracks resolution of these "generic safety issues." The generic safety issue program provides for (1) identifying generic issues, (2) assigning them priorities, (3) developing detailed action plans for their resolution, (4) overseeing progress in their resolution by senior managers, and (5) informing the public of the status of progress in resolution. The resolution of these issues may involve new or revised rules, new or revised guidance, or revised interpretation of rules or guidance that affect nuclear power plant licensees or nuclear material certificate holders. The U.S. Congress requires that the NRC maintain this program.

The NRC promulgated the License Renewal Rule in 1995 (in 10 CFR Part 54). The license renewal process focuses on passive and long-lived SSCs because degradation in active components is more readily detected by complying with the Maintenance Rule. License renewal applicants are required to complete an environmental assessment and an integrated plant assessment[6] and to evaluate time-limited aging analyses. The current licensing basis must be maintained throughout the period of extended operation. (Section 14.1.2 of this report describes the NRC license renewal process.)

14.1.3.5 Risk-Informed Regulation and the Reactor Oversight Process

The NRC is actively increasing the use of risk insights and information in its regulatory decisionmaking. For reactors, risk-informed activities occur in the five broad categories of (1) applicable regulations, (2) licensing process, (3) Reactor Oversight Process, (4) regulatory guidance, and (5) risk analysis tools, methods, and data. Activities within these categories include revisions to technical requirements in the regulations; risk-informed technical specifications; a framework for inspection, assessment, and enforcement actions; guidance on risk-informed inservice inspections; and improved standardized plant analysis risk models.

In 2000, the NRC implemented a revised Reactor Oversight Process using risk insights and lessons learned from more than 40 years of regulating nuclear power plants. The previous oversight process evolved during a period when the nuclear power industry was less mature and there was much less operational experience on which to base rules and regulations. Very conservative judgments governed the rules and regulations. Significant plant operating events occurred with some frequency, and the oversight process tended to be reactive and prescriptive, closely observing plant performance for adherence to the regulations and responding to operational problems as they occurred.

After nearly four decades of operational experience and generally steady improvements in plant performance, the Reactor Oversight Process now focuses more of the agency's resources on the relatively small number of plants with performance problems. The process is a means to collect information about licensee performance, assess the information for its safety significance, and provide for appropriate licensee and NRC response, including corrective and enforcement actions, when appropriate. Areas such as emergency preparedness, radiation safety, human performance, safety culture, and problem identification and resolution are among those evaluated.

The Reactor Oversight Process makes greater use of objective performance indicators. Together, the performance indicators and inspection findings give the information needed to support quarterly reviews of plant performance. The Reactor Oversight Process also features expanded semiannual reviews, which include inspection planning and a performance report (all posted on the NRC's public Web site). The Reactor Oversight Process is more effective at correcting performance or equipment problems today because the agency's response to

---

[6]   An integrated plant assessment identifies and lists structures and components subject to an aging management review. These include "passive" structures and components that perform their intended function without moving parts or without a change in configuration or properties. Examples of these are the reactor vessel, the steam generators, piping, component supports, and seismic Category I structures. To be in scope, the item must also be long-lived to be considered during the license renewal process. Long-lived means the item is not subject to replacement based on a qualified life or specified time period.

problems is more timely and predictable.   (Section 6.3.2 of this report provides a full description of the NRC Reactor Oversight Process.)

14.1.3.6   Licensee Responsibilities for Safety: Regulations and Initiatives Beyond Regulations

As in many countries, U.S. nuclear power plant licensees are responsible for the safety of their facilities.   This responsibility is embedded in their license and in the NRC's regulatory infrastructure.   Under the regulatory umbrella, licensees routinely assess new technologies, off-normal conditions, operating experience, and industry trends to make informed decisions about safety enhancements to their facilities.

Under the U.S. regulatory structure, 10 CFR Part 50, Appendix B requires that all nuclear power plant licensees maintain a quality assurance program.   Quality assurance comprises all those planned and systematic actions necessary for adequate confidence that an SSC will perform satisfactorily in service.   Quality assurance includes quality control, which comprises those quality assurance actions related to the physical characteristics of a material, structure, component, or system that provide a means to control quality to predetermined requirements.

Licensees carry out a comprehensive system of planned and periodic audits to verify compliance with all aspects of the quality assurance program and to determine the effectiveness of the program.   Appropriately trained personnel who do not have direct responsibilities in the areas being audited perform these audits in accordance with written procedures or checklists.   Audit results are documented and reviewed by management with responsibility in the area audited, and appropriate followup is initiated.

14.1.3.7   The NRC's Regulatory Process Compared with International Safety Reviews

IAEA and the Western European Nuclear Regulators' Association (WENRA) have developed guidance[7] and objectives for conducting periodic safety reviews that have much in common. Consistent with the guidance of both organizations, periodic safety reviews are comprehensive assessments with the following purposes:

- to determine, at the time of the review, whether the plant complies with its licensing basis

- to identify the extent to which the current licensing basis remains valid, in part by determining the extent to which the plant meets current safety standards and practices

- to provide a basis for implementing appropriate safety improvements, corrective actions, or process improvements

- to provide confidence that the plant can continue to be operated safely

---

[7]   IAEA guidance appears in Safety Standards Series No. NS-G-2.10, "Periodic Safety Review of Nuclear Power Plants Safety Guide," issued in 2003.   WENRA guidance appears in "Pilot Study on Harmonization of Reactor Safety in WENRA Countries," WENRA Working Group on Reactor Harmonization, March 2003.

For the reasons discussed above and summarized below, the shared objectives associated with the IAEA and WENRA periodic safety review guidance are substantively accomplished in the United States on an ongoing basis.

First, the NRC's regulatory process provides a robust foundation for ongoing assessments, evaluations, and, when appropriate, imposition of new requirements. Currently, the NRC and the U.S. nuclear industry consider new information in a more risk-informed manner as it becomes available; adjust the regulatory oversight and plant safety priority, respectively; and provide ongoing assurance that the licensing basis for the design and operation of all nuclear power plants provides an acceptable level of safety. Development of the Maintenance Rule and License Renewal Rule are two examples of new requirements that serve this purpose.

Second, the NRC and the U.S. nuclear industry have a 30-year history of implementing broad-based plant assessments. The regulatory history of implementing broad-based assessments is a direct result of an adaptive, probing, and independent regulatory process. These assessments have included the SEP, the ISAP, and the individual plant examinations. They provide additional confidence that plant safety continues to be the highest priority and that the NRC and industry continue to pursue enhancements that improve safety. As shown in the figure included below, over a period of almost 25 years, broad-based NRC assessments and regulatory initiatives have provided a continuum of assessment, improvement, and oversight, which ensures that licensed plants continue to operate safely.

The NRC's transition to a more risk-informed regulatory framework and the Reactor Oversight Process offers an ongoing approach and basis for implementing appropriate safety improvements, corrective actions, or process improvements and provides confidence that the plant can continue to be operated safely. The NRC's more risk-informed approach helps ensure that resources are optimally focused on those issues most important to safety.

Finally, U.S. licensees establish performance expectations above the thresholds required by the NRC. These self-imposed expectations and initiatives -- over and above the regulations – result from the licensee's self-described motivation to pursue excellence and by the recognition that safety and economics are directly linked in the competitive, free-market U.S. energy industry.

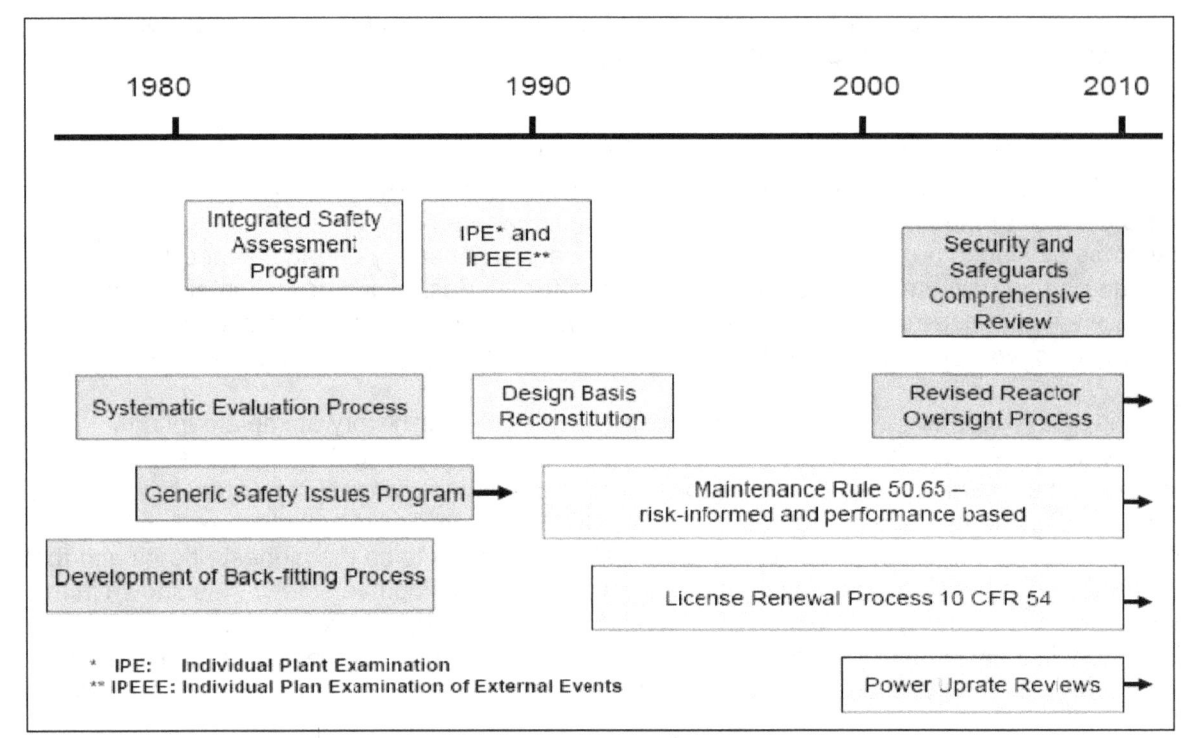

14.2   Verification by Analysis, Surveillance, Testing, and Inspection

Licensees are required to verify that they are operating their nuclear installations in accordance with the plant-specific design and requirements.   The technical specifications (for surveillance) and national consensus codes (for testing and periodic inspections) contain the requirements for verification.

In 10 CFR 50.55a, "Codes and Standards," the NRC gives requirements for applying industry codes and standards to nuclear power reactors during design, construction, and operation. This section states, "Systems and components of boiling and pressurized water-cooled nuclear power reactors must meet the requirements of the ASME Boiler and Pressure Vessel Code specified in paragraphs (b) through (g) of this section."   In addition, 10 CFR 50.55a provides for alternatives to the ASME Code when authorized by the NRC.

Through analysis, surveillance, testing, and inspection, the NRC verifies that the physical state and operation of nuclear installations continue to be in accordance with the designs, applicable national safety requirements, and operational limits and conditions.   As discussed in Article 6 of this report, the NRC's Reactor Oversight Process includes inspections to verify that licensees are fulfilling their obligations to carry out such surveillances and testing and take corrective action. The agency's Reactor Oversight Process collects the data for performance measure in two ways. First, NRC inspectors collect inspection findings at least quarterly, using formal detailed IPs to review plant operations and maintenance.   NRC managers review inspection findings to assess their significance as part of the Reactor Oversight Process' significance determination process. Second, licensees collect data for performance indicators and submit this information to the NRC at least quarterly.   The thresholds for each indicator determine the significance of the data. The NRC performs inspections of licensee processes for collecting and submitting the data to

124

ensure completeness, accuracy, consistency, timeliness, and validity. The NRC publishes the inspection findings and performance indicators on its Web site and incorporates feedback from all stakeholders as appropriate.

Annually, senior agency managers review plants that have performance issues and report these results to the Commission. An integral part of the evaluative process used by the agency to ensure the operational safety performance of nuclear licensees, this annual Agency Action Review Meeting provides another opportunity for the NRC's senior management to discuss significant events, licensee performance issues, trends, and the actions to mitigate recurrences.

The NRC also focuses on aging management. NUREG-1800, Chapter 3, "Aging Management Review Results," dated September 2005, addresses aging management review of reactor vessel, internals, reactor coolant system, engineered safety features, auxiliary systems, steam and power conversion system, containment, structures, component supports, electrical systems and instrumentation and controls. NUREG-1800, Chapter 4, "Time-Limited Aging Analyses," addresses the identification of time-limited aging analyses. The list of potential time-limited aging analyses comprises certain plant-specific safety analyses that are based on an assumed 40-year plant life. Under 10 CFR 54.21(c)(1), the NRC requires a license renewal applicant to list time-limited aging analyses, as defined in 10 CFR 54.3, "Definitions." The NRC evaluates the adequacy of the time-limited aging analyses identified by the applicant.

Under special circumstances, the Commission may also require under 10 CFR 50.54(f) that licensees submit written statements to enable the Commission to determine whether the license should be modified, suspended, or revoked.

The NRC updates, revises, and improves existing regulatory programs in light of operating experience and significant new safety information. Article 19 of this report discusses these activities.

ARTICLE 15.   RADIATION PROTECTION

Each Contracting Party shall take the appropriate steps to ensure that, in all operational states, the radiation exposure to the workers and to the public caused by a nuclear installation shall be kept as low as reasonably achievable, and that no individual shall be exposed to radiation doses which exceed the prescribed national dose limits.

This section summarizes the authorities and principles of radiation protection, which include the regulatory framework, regulations, and radiation protection programs for controlling radiation exposure for occupational workers and members of the public.   Article 17 of this report discusses radiological assessments that apply to licensing and facility changes.

15.1   Authorities and Principles

Generally, U.S. radiation control measures are founded on radiological risk assessments by the United Nations Scientific Committee on the Effects of Atomic Radiation and the U.S. National Academy of Sciences Committee on the Biological Effects of Ionizing Radiation.   The risk management recommendations promulgated by the International Commission on Radiological Protection (ICRP) and the National Council on Radiation Protection and Measurements (NCRP) reflect these assessments.   On the basis of these assessments and recommendations, EPA develops Federal guidance signed by the President of the United States, and "generally applicable radiation standards" for use by the other Federal agencies, including the NRC.   The responsible agencies, such as the NRC, then establish regulations that consider these recommendations and standards.   U.S. radiation protection programs are based on principles generally consistent with the principles espoused by ICRP: (1) it is known that large doses of ionizing radiation can be deleterious to human health, and (2) it is considered prudent to assume that small doses may also be harmful, with the probability of a deleterious effect being proportional to the dose.   The U.S. programs acknowledge, include, and use the ICRP-recommended protection principles of "limitation," "justification," and "optimization" as appropriate.

Of these principles, "limitation" is the most practicable and most directly included in the regulatory structure.   The regulations establish dose limits that cannot be exceeded without violating the regulations.   There is a lengthy history of the doses being kept within the limits for workers (NUREG-0713, Volume 30, "Occupational Radiation Exposure at Commercial Nuclear Power Reactors and Other Facilities," dated January 2010) and members of the public living near nuclear power plants (NUREG/CR-2850, Volume 14, "Dose Commitments Due to Radioactive Releases from Nuclear Power Plant Sites in 1992," dated March 1996).

"Justification," is the recommendation that any activity involving radiation exposure be shown to be beneficial before the activity is undertaken.   However, the risks or benefits of a new application of radioactive material can seldom be determined in advance with complete accuracy. Furthermore, radiation protection considerations are only one contributor to overall decisions on whether a particular exposure situation is justified.   The "justification" activities in the U.S. are carried out during the licensing process.   In general, the NRC will reject an application to use or produce radioactive materials if it determines that the application is frivolous (i.e., that the overall benefit to society is outweighed by the risk of the radiation exposure associated with the activity). For some large applications, such as the generation of electricity from nuclear power, national policy establishes the justification.   Because national energy policy favors nuclear power (i.e., the net benefit for the United States is deemed to be positive), the licensing process under

10 CFR Part 50 does not specifically address the justification for licensing a nuclear power plant.

Rather than using the term "optimization," the U.S. has used the term "as low as is reasonably achievable" (ALARA). In most circumstances, these two terms are consistent and represent the same underlying principle. As a guiding principle, ALARA (with varying terminology) dates back to 1939 in the U.S. and is defined in the regulations for occupational workers and members of the public.

For decades before 1994, 10 CFR Part 20 addressed the ALARA criterion for occupational radiation exposure, but more as an admonition than as a requirement. In 1994, the NRC changed the regulation to require that all licensees develop, document, and carry out an ALARA program. The NRC would judge compliance with this requirement on the basis of a licensee's capability to track and, if necessary, reduce exposures, rather than on whether exposures and doses represented an absolute minimum or whether the licensee had used all possible methods to reduce exposures.

For control of radiation exposure from nuclear power plants to members of the public, the NRC modified 10 CFR Part 50 by adding Appendix I, "Numerical Guides for Design Objectives and Limiting Conditions for Operation to Meet the Criterion "As Low As Is Reasonably Achievable" for Radioactive Material in Light-Water-Cooled Nuclear Power Reactor Effluents." Issued in 1975, this appendix established design objectives to keep radioactive releases from nuclear power plants ALARA. The ALARA requirement led to the establishment of numerical objectives (for example, 0.00005 sievert (Sv) (0.005 rem) in a year for the most highly exposed individual). Similar EPA requirements for other facilities soon followed. These NRC and EPA requirements are consistent with ICRP principles and result in public doses that are well below the local variation in doses from natural sources.

Although U.S. regulations are generally consistent with ICRP recommendations, certain constraints have limited the extent to which U.S. regulations match those of ICRP. One important constraint has been the U.S. desire for regulatory stability. Revising the regulations to incorporate every new ICRP position would impose a serious burden on the licensees without a commensurate benefit. Furthermore, for nuclear power reactors, new requirements are constrained by the Backfit Rule's requirements that any increase in regulatory requirements other than those required for compliance with existing regulations or the statutory standard of "adequate protection" be justified by a commensurate improvement in safety (see 10 CFR 50.109). Consequently, U.S. regulations were founded on older (rather than the most recent) ICRP recommendations. Nevertheless, the NRC directed the staff to work closely with ICRP and other national and international organizations to help develop revised recommendations. After publication of the new ICRP recommendations (ICRP Publication 103, "The 2007 Recommendations of the International Commission on Radiological Protection," dated March 2007), the NRC staff provided options for Commission consideration in SECY-08-0197, "Options to Revise Radiation Protection Regulations and Guidance with Respect to the 2007 Recommendations of the International Commission on Radiological Protection," dated December 18, 2008. The Commission approved the staff initiating stakeholder dialogue and technical basis development to explore the benefits and effects of increasing alignment with ICRP. As part of this process, the NRC staff is currently in active dialogue with all segments of the licensed community in the U.S. The NRC may revise its regulations, in whole or in part, depending on the outcome of these discussions.

15.2   Regulatory Framework

The NRC developed requirements for radiation protection to implement three laws passed by the U.S. Congress: the Atomic Energy Act of 1954, as amended; the Energy Reorganization Act of 1974; and the Uranium Mill Tailings Radiation Control Act of 1978.

NRC regulations establish the primary direct controls over licensees.   Various documents provide additional guidance and clarification, including RGs, topical staff and contractor reports (NUREG series), GLs, technical specifications, and license conditions.   These documents are supported by international standards, consensus national standards, and authoritative recommendations (such as those of ICRP and NCRP).   However, these supporting documents have no official status unless they are referenced in or adopted by a regulation or documents providing regulatory guidance, such as RGs or standard review plans.   Of particular importance are NUREG-0800, which guides the staff in reviewing safety analysis reports, and RG 1.70, Revision 3, "Standard Format and Content of Safety Analysis Reports for Nuclear Power Plants," dated November 1978, which guides the applicant in writing safety analyses.   Chapter 11 of NUREG-0800 addresses the control of radioactive effluents.   Chapter 12 addresses radiation protection.   Chapter 15 details how to calculate offsite and control room operator doses for design-basis accidents.   Under 10 CFR 50.34(g), the facility must be evaluated against the standard review plan.

As Article 6 of this report discussed, the Reactor Oversight Process has cornerstones for radiation safety.   The cornerstone public radiation safety focuses on the effectiveness of the plant's programs in meeting applicable Federal limits on the exposure, or potential exposure, of members of the public to radiation and in ensuring that the effluent releases from the plant are ALARA.   The cornerstone for occupational radiation safety focuses on the effectiveness of the plant's program(s) in maintaining the worker dose within the regulatory limits and providing occupational exposures that are ALARA.

15.3   Regulations

The regulations that apply to radiation protection are 10 CFR Part 20 and 10 CFR Part 50.

10 CFR Part 20.   The NRC regulations in 10 CFR Part 20 establish requirements for radiation protection for all NRC licensees.   The NRC gives additional requirements for specific operations and specific kinds of licenses in other parts of Title 10:   10 CFR Part 30, "Rules of General Applicability to Domestic Licensing of Byproduct Material;" 10 CFR Part 34, "Licenses for Industrial Radiography and Radiation Safety Requirements for Industrial Radiographic Operations;" 10 CFR Part 35, "Medical Use of Byproduct Material;" 10 CFR Part 39, "Licenses and Radiation Safety Requirements for Well Logging;" 10 CFR Part 40, "Domestic Licensing of Source Material;" 10 CFR Part 50; 10 CFR Part 70, "Domestic Licensing of Special Nuclear Material;" 10 CFR Part 71, "Packaging and Transportation of Radioactive Material;" and 10 CFR Part 72.

The most recent major revision of 10 CFR Part 20, issued in 1991, adopted the recommendations, quantities, and models recommended in ICRP Publication 26, "Recommendations of the International Commission on Radiological Protection," dated January 1977, and in ICRP Publication 30, "Limits of Intakes of Radionuclides by Workers," dated 1978-1982, as well as some recommendations from NCRP Report No. 91, "Recommendations on Limits for Exposure to Ionizing Radiation," dated June 1987.   The 1991 revision to 10 CFR Part 20 also adopted the same dose limit for a member of the public recommended in ICRP Publication 60, "1990 Recommendations of the International Commission on Radiological Protection" dated November 1990. Providing relatively comprehensive coverage of general requirements for radiation protection, 10 CFR Part 20 is divided into subparts, with each subpart addressing a specific area of radiation protection, such as occupational and public dose limits, posting, surveys, monitoring, waste disposal, and reporting.

The details of the requirements in 10 CFR Part 20 are not entirely consistent with international standards such as IAEA's Safety Standards, Safety Series No. 115, "International Basic Safety Standards for Protection against Ionizing Radiation and for the Safety of Radiation Sources," dated February 1996. The main areas of difference between 10 CFR Part 20 and the Basic Safety Standards include the use of the effective dose equivalent in 10 CFR Part 20 versus use of the effective dose in the Basic Safety Standards, an annual occupational dose limit on the effective dose equivalent of 0.05 Sv in 10 CFR Part 20 versus 0.02 Sv in the Basic Safety Standards, and use of the biokinetic models from ICRP Publication 30 in 10 CFR Part 20 versus the more recent models used in the Basic Safety Standards. The NRC is engaging stakeholders in a dialogue to consider revising its regulations in the near future to better align with new international standards. In the interim, NRC licensees are permitted to use the effective dose in place of the effective dose equivalent and to use the more recent internal dosimetry models in place of those recommended in ICRP Publication 30, with prior NRC approval.

In addition, many licensees and agencies have administrative dose limits that are similar to or lower than those in the Basic Safety Standards. Most other licensees operate at occupational doses far below those limits and standards and therefore are considered ALARA. In some cases, the occupational doses do exceed 0.02 Sv per year (2 rem per year), but these are a very small fraction of the total, and efforts are continuing to reduce these doses to lower levels. In the interim and until the completion of NRC considerations that may better align its regulations with international standards, the current 10 CFR Part 20 provides a level of radiation protection that in almost all situations is comparable to that provided by international standards.

10 CFR Part 50. Although 10 CRR Part 50 is the principal regulation addressing the safety of nuclear power plants, only a small section of it directly addresses radiation protection. Even so, the sections of 10 CFR Part 50 that do affect radiation protection are significant. Of particular importance are 10 CFR 50.34a, "Design Objectives for Equipment to Control Releases of Radioactive Material in Effluents-Nuclear Power Reactors," 10 CFR Part 50, Appendix I, "Numerical Guides for Design Objectives and Limiting Conditions for Operation to Meet the Criterion "As Low as is Reasonably Achievable" for Radioactive Material in Light-Water-Cooled Nuclear Power Reactor Effluents," and 10 CFR 50.34(g), which requires NRC review of in-plant radiation protection program. In 10 CFR 50.36a, "Technical Specifications on Effluents from Nuclear Power Reactors," the NRC also requires licensees to limit effluents from nuclear power reactors to the values in Appendix I to 10 CFR Part 50. The revised dose criteria for design-basis accidents appear in 10 CFR 50.34(a)(1)(ii)(D) for licensing actions after implementation of the revised rule in 1997. (The dose criteria for siting and determining the exclusion area low population zone and population center distance for nuclear power reactors appear in 10 CFR 100.11(a).)

15.4 Radiation Protection Activities

Radiation protection activities apply to occupational workers and to members of the public.

15.4.1 Control of Radiation Exposure of Occupational Workers

In addition to focusing on personnel qualifications for licensing, the NRC's oversight and regulation of radiation protection programs ensure that the safety analysis report and radiation protection plan properly address each item in 10 CFR Part 20, as well as the provisions for instructions to workers in 10 CFR Part 19, "Notices, Instructions, and Reports to Workers: Inspection and Investigations," and the provisions in relevant RGs, such as RG 1.8, "Qualification and Training of Personnel for Nuclear Power Plants," dated March 1971 (as revised September 1975, May 1977, April 1987, and May 2000), and RG 8.8, Revision 3, "Information Relevant to Ensuring that Occupational Radiation Exposures at Nuclear Power Stations Will Be As Low As Is Reasonably Achievable," dated June 1978.

Once the NRC issues a license, it maintains an active regulatory program that includes routine inspection and monitoring of nuclear plants to alert NRC staff of potential problems in radiation safety.   Significant health physics problems can trigger significant reactive regional inspections or a generic communication to the industry.

The NRC staff has been collecting the annual occupational exposure data for light-water reactors since 1969.   Because the amount and kind of maintenance performed strongly influences the doses, the individual plant collective doses fluctuate from year to year.   Still, clear trends are evident.   Using the average collective dose per reactor as the reference, statistical analysis shows that the doses varied almost randomly before the accident at Three Mile Island Unit 2. Thereafter, the doses increased as a result of the extensive modifications required of all nuclear power plants in response to new NRC requirements.   The average collective dose reached a peak of 7.91 person-Sv (791 person-rem) per reactor in 1980.   Since then, doses have declined almost steadily to the current level below 1 person-Sv (100 person-rem) per reactor, where they have remained for the past 5 years (2004–2008, the last year for which the data have been compiled).   The 2008 average collective dose value of 0.88 person-Sv (88 person-rem) per reactor was the lowest average collective dose recorded since data collection began in 1969. Although the average doses for both PWRs and BWRs have been steadily declining, the average BWR dose has exceeded the average PWR dose since 1974.   Over the past 5 years, the average BWR dose has exceeded the average PWR dose by roughly 90 percent (in part because of the higher average dose rates and larger work force at BWRs).
In 2008, the 118,692 workers at nuclear plants received 91.96 person-Sv (9,196 person-rem) for an average of 0.00077 Sv (0.077 rem) per worker.   This represents a 92-percent drop in average worker dose from the 1973 value of 0.0095 Sv (0.95 rem) per worker.

15.4.2 Control of Radiation Exposure of Members of the Public

The regulations in 10 CFR 20.1301, "Dose Limits for Individual Members of the Public," and 10 CFR 20.1302, "Compliance with Dose Limits for Individual Members of the Public," control radiation exposures to members of the public.   In addition to the 1.0 millisievert (100 millirem) annual dose limit in 10 CFR Part 20, the EPA regulations in 40 CFR Part 190, "Environmental Radiation Protection Standards for Nuclear Power Operations," establish a regulatory standard such that the annual dose to a member of the public from exposures to sources associated with the entire uranium fuel cycle does not exceed 0.25 millisievert (25 millirem).

The regulations in 10 CFR 50.34a, 10 CFR 50.36a, and 10 CFR Part 50, Appendix I, define the ALARA plant objectives for effluents. Appendix I also specifies effluent monitoring, environmental monitoring, investigations, land-use censuses, and reporting. Section IV.B of 10 CFR Part 50, Appendix I, requires the licensee to establish an appropriate surveillance and monitoring program that will accomplish the following:

- Provide data on quantities of radioactive material released in liquid and gaseous effluents.

- Provide data on measurable levels of radiation and radioactive materials in the environment to evaluate the relationship between quantities of radioactive material released in effluents and resultant radiation doses to individuals from principal pathways of exposure.

- Identify changes in the use of unrestricted areas (e.g., for agricultural purposes) to permit modifications in monitoring programs for evaluating doses to individuals from principal pathways of exposure.

Appendix I requirements are supplemented by 10 CFR Part 20.1501, "General," which requires, in part, that a licensee perform surveys to evaluate potential radiological hazards and to demonstrate compliance with the public dose limits in 10 CFR 20.1301 and 10 CFR 20.1302. Therefore, a licensee is responsible for performing radiation surveys at its facility for radioactive materials that have the potential to affect workers and members of the public. Potential survey sites can include areas that have been previously affected by licensed radioactive material, as well as areas that may be affected by licensed radioactive material in the future. For onsite spills and leaks that may contain licensed radioactive material, 10 CFR 20.1501 requires a licensee to perform appropriate radiation surveys and monitoring to determine the radiological hazard (i.e., dose assessment) to workers and to determine if there is a viable pathway to the unrestricted area that could result in a potential radiological hazard to members of the public. The surveys and monitoring can continue over a period of time or become an ongoing monitoring program so that the licensee can adequately characterize the extent and source of the contamination from the spills or leak

Since 2004, there have been several discoveries of radioactive ground water contamination at nuclear power facilities in the U.S. Investigation has determined that most of the contamination resulted from undetected leakage from facility SSCs that contained or transported radioactive liquids. All unmonitored releases resulted in varying levels of onsite tritium ground water contamination, with two facilities detecting low levels of tritium (below EPA drinking water standards) in offsite residential drinking wells. Current data show no immediate public health effects and a very low probability that there will be an effect in the future.

The NRC has responded to reports of ground water contamination by carrying out inspections, assessing the safety significance of these events, and evaluating licensee performance in finding and taking corrective actions. The NRC has also issued INs 2004-05, "Spent Fuel Pool Leakage to Onsite Groundwater," dated March 3, 2004, and IN 2006-13, "Ground-Water Contamination Due to Undetected Leakage of Radioactive Water," dated July 10, 2006, describing unmonitored and unplanned leakage at several nuclear power stations.

Both the NRC and the nuclear industry have worked to resolve the technical and programmatic issues leading to the ground water contamination events. In March 2006, the NRC Executive Director for Operations established a Liquid Radioactive Release Lessons Learned Task Force to assess lessons learned from the unmonitored release of radioactive liquid to the environment at power reactor sites and to recommend possible agency actions.   The task force completed its assessment and issued its report on September 1, 2006.   The most significant conclusion was that these events had no public health effect.   However, because of the high level of public concern and the potential for contaminated ground water to migrate off site undetected, the task force made several recommendations to the NRC.   In response to the task force recommendations, the NRC revised its guidance in RG 1.21, "Measuring, Evaluating, and Reporting Radioactive Material in Liquid and Gaseous Effluents and Solid Waste," and RG 4.1, "Radiological Environmental Monitoring for Nuclear Power Plants," both dated June 2009, to clarify its expectations concerning monitoring and reporting leaks and spills.

In parallel with the NRC's efforts, the nuclear industry also responded to the ground water contamination events.   The NEI has developed a voluntary Groundwater Protection Initiative that licensees have endorsed unanimously.   The initiative required each participating nuclear plant to have a plan in place by July 2006 that established several short-term actions, such as developing an enhanced communications protocol to ensure notification of State and local officials of less significant unmonitored release events.   The industry initiative also required several long-term actions to improve leak detection monitoring capability and understanding of site hydrology and geology.

The NRC has initiated a special inspection effort to monitor the licensee's implementation of the industry's Groundwater Protection Initiative.   As a result of the enhanced monitoring, the NRC has identified several additional occurrences of low-level tritium contamination in onsite ground water.   To date, levels of contamination have been below any NRC-required reporting level and well below the ALARA dose objectives in 10 CFR 50, Appendix I.   However, the NRC continues to oversee licensee's responses to each of these occurrences and is actively considering whether additional regulatory requirements or guidance are warranted for the integrity of buried piping and subsurface SSCs.

In addition, in March 2010 the NRC established a task force to evaluate its regulatory framework associated with groundwater protection.   The objective of the task force was to evaluate NRC actions to date addressing buried piping leaks and whether those actions needed to be augmented.   The report "Groundwater Task Force Final Report," dated June 2010, documents the task force's observations, conclusions, and recommendations in a number of areas, including policy and communications.   Currently, a Senior Management Review Group has been formed from a group of NRC senior executives, and has been tasked to decide whether the NRC agrees with the findings of the task force and how best to act upon the conclusions and recommendations contained in the final report.

ARTICLE 16. EMERGENCY PREPAREDNESS

(i)   Each Contracting Party shall take the appropriate steps to ensure that there are onsite and offsite emergency plans that are routinely tested for nuclear installations, and cover the activities to be carried out in the event of an emergency.

(ii)   For any new nuclear installation, such plans shall be prepared and tested before it [the installation] commences operation above a low power level agreed [to] by the regulatory body.

(iii)   Each Contracting Party shall take appropriate steps to ensure that, insofar as they are likely to be affected by a radiological emergency, its own population and the competent authorities of the States in the vicinity of the nuclear installation are provided with appropriate information for emergency planning and response.

(iv)   Contracting Parties that do not have a nuclear installation on their territory, insofar as they are likely to be affected in the event of a radiological emergency at a nuclear installation in the vicinity, shall take the appropriate steps for the preparation and testing of emergency plans for their territory that cover the activities to be carried out in the event of such an emergency.

This section discusses (1) emergency planning and emergency planning zones, (2) offsite emergency planning and preparedness, (3) emergency classification system and action levels, (4) recommendations for protection in severe accidents, (5) inspection practices and regulatory oversight, (6) response to an emergency, and (7) international arrangements.

16.1 Background

The NRC's responsibilities for radiological emergency preparedness stem from NRC licensing functions under the Atomic Energy Act and the Energy Reorganization Act.   Both statutes authorize the Commission to promulgate regulations that it deems necessary to fulfill its responsibilities under the acts.   Following the accident at Three Mile Island Unit 2 in March 1979, the NRC amended the regulations to require significant changes in emergency planning and preparedness for U.S. commercial nuclear power plants.   The NRC's emergency planning regulations are now an important part of the regulatory framework for protecting public health and safety and have been adopted as an added conservatism in the NRC's defense-in-depth safety philosophy of multiple-barrier containment and redundant safety systems.   Before a full-power operating license can be issued, NRC regulations require a finding that there is reasonable assurance that adequate measures to protect public health and safety can and will be taken in a radiological emergency (10 CFR 50.47(a)).

Emergency planning in the United States recognizes that a spectrum of accidents could exceed the design-basis accidents that nuclear plants are required to accommodate without significant public health and safety effects.   For design-basis accidents, the small releases that might occur would not likely require responses such as evacuating or sheltering the general public.   These actions become important only when considering accidents that are much less probable than design-basis accidents.   NUREG-0396, "Planning Basis for the Development of State and Local Government Radiological Emergency Response Plans in Support of Light-Water Nuclear Power Plants," dated December 1978, and NUREG-0654/FEMA-REP-1 (NUREG-0654),

Revision 1, "Criteria for Preparation and Evaluation of Radiological Emergency Response Plans and Preparedness in Support of Nuclear Power Plants," dated November 1980, describe the emergency planning basis.

16.2 Offsite Emergency Planning and Preparedness

The accident at Three Mile Island Unit 2 revealed that better coordination and more comprehensive emergency plans and procedures were needed if the NRC and the public were to have confidence in the readiness of onsite and offsite emergency response organizations to respond to a nuclear emergency.   Participation by State and local governments in emergency planning for nuclear power plants in the United States was, and still remains, largely voluntary.   Before the accident at Three Mile Island Unit 2, there had been no clear obligation for State and local governments to develop emergency plans for radiological accidents, and the Federal role was one of assistance and guidance.   After the accident, the NRC amended its emergency planning regulations to require, as a condition of licensing, that each applicant or licensee submit the radiological emergency response plans of the State and local governments that are within the plume exposure zone, as well as the plans of State governments within the ingestion pathway zone (10 CFR 50.33(g) and 10 CFR 50.54(s)).

In December 1979, the President directed FEMA to take the lead in ensuring the development of acceptable State and local offsite emergency plans and activities for nuclear power plants.   The NRC and FEMA regulations, as well as a memorandum of understanding between the two agencies, dated June 17, 1993, subsequently codified the role and responsibilities of FEMA.

FEMA provides its findings on the acceptability of the offsite emergency plans to the NRC, which has the ultimate responsibility for determining the overall acceptability of radiological emergency plans and preparedness for a nuclear power reactor.   The NRC will not issue a license to operate a nuclear power reactor unless it finds that the state of onsite and offsite emergency preparedness provides reasonable assurance that protective measures can and will be taken in a radiological emergency.   The NRC bases its decision on a review of the FEMA findings and determinations on whether State and local emergency plans are adequate and can be carried out, and on its own assessment of whether the onsite emergency plans are adequate and can be implemented (10 CFR 50.47(a)).

The principal guidance for preparing and evaluating radiological emergency plans for licensee, State, and local government emergency planners is NUREG-0654, a joint NRC and FEMA document.   NUREG-0654 gives evaluation criteria for meeting the emergency planning standards in the NRC and FEMA regulations (10 CFR 50.47(b) and 44 CFR Part 350, "Review and Approval of State and Local Radiological Emergency Plans and Preparedness," respectively).   These criteria provide a basis for licensees, States, and local governments to develop acceptable emergency plans.

The NRC and FEMA coordinate their evaluation of periodic emergency response exercises and require all operating nuclear power plant sites to conduct an exercise every 2 years, as outlined in Section IV.F.2(b) of 10 CFR Part 50, Appendix E.   These mandatory full-participation exercises are integrated efforts by the licensee, State, and local radiological emergency response organizations that have a role in support of the licensee's emergency plan.   The NRC evaluates the licensee's performance, while FEMA evaluates the response by State and local agencies.   In some cases, other Federal response agencies also participate in these exercises.   Any weaknesses or deficiencies identified by the NRC or FEMA as a result of the exercise must be corrected through appropriate remedial actions.   Section IV.F.2(d) of Appendix E, requires the State's response personnel to participate in biennial exercises of their plume exposure pathway plans every 2 years and in an ingestion pathway exercise with a nuclear power plant located within their State every 6 years. However, there are no requirements to involve members of the public in any of the emergency preparedness exercises.

16.3 Emergency Classification System and Emergency Action Levels

NRC regulations establish four classes of emergencies in order of increasing severity: (1) unusual event, (2) alert, (3) site area emergency, and (4) general emergency. The specific class of emergency is declared on the basis of plant conditions that trigger the emergency action levels. Licensees have established specific procedures for carrying out emergency plans for each class of emergency. The event classification initiates all appropriate actions for that class, including notification of offsite authorities, activation of onsite and offsite emergency response organizations, and -- where appropriate -- protective action recommendations for the public. These same emergency classes are also found in the State and local emergency plans that support each nuclear power plant.

NUREG-0654 gives examples of initiating conditions for each of the four emergency classes. These conditions form the basis for each licensee to establish specific indicators, known as "emergency action levels." These levels provide a clear basis for rapidly identifying a possible problem, alerting the onsite emergency response organization, and notifying the offsite authorities that an emergency exists. NRC regulations require the licensee, State, and local government authorities to discuss and agree upon the emergency classification levels, which the NRC must approve. In RG 1.101, Revision 4, "Emergency Planning and Preparedness for Nuclear Power Reactors," dated July 2003, the NRC endorsed the guidance in NUMARC/NESP-007, Revision 2, "Methodology for Development of Emergency Action Levels," dated January 1992; and NEI 99-01, Revision 4, "Methodology for Development of Emergency Action Levels," dated January 2003, as acceptable alternatives for developing emergency action levels.

16.4 Recommendations for Protective Action in Severe Accidents

The technical basis and guidance for determining protective actions in the United States for severe (core damage) reactor accidents appear in NUREG-0654/FEMA-REP-1, Supplement 3, Revision 1, "Criteria for Protective Action Recommendations for Severe Accidents," dated July 1996, and EPA 400-R-92-001, "Manual of Protective Action Guides and Protective Actions for Nuclear Incidents," dated May 1992. These documents reflect the conclusions that were developed from severe accident studies, such as NUREG-1150, "Severe Accident Risks: An Assessment for Five U.S. Nuclear Power Plants," dated December 1990.

The agency provides guidance for response procedures and training manuals for NRC staff in NUREG/BR-0150, Volume 1, Revision 4, "Response Technical Manual 96," dated March 1996. The NRC's guidance on evacuation and sheltering in the event of a nuclear power plant accident is consistent with guidance in IAEA TECDOC-953, "Method for the Development of Emergency Response Preparedness for Nuclear or Radiological Accidents," and TECDOC-955, "Generic Assessment Procedures for Determining Protective Actions During a Reactor Accident," both issued in 1997.

The NRC considers evacuation and sheltering to be the two primary protective actions and prefers prompt evacuation for the population near a plant in a severe reactor accident. However, the NRC is currently evaluating this position, as under some circumstances, it may be better to shelter in place.   A draft revision to NUREG-0654/FEMA-REP-1, Supplement 3, was published in the *Federal Register* on March 8, 2010, for public comment and placed on http://www.regulations.gov under docket NRC-2010-0080.

A supplemental protective action for the general population is using the thyroid-blocking agent potassium iodide.   The NRC amended its regulations for emergency planning associated with potassium iodide, 10 CFR 50.47(b)(10), in 2001.   This amendment requires that each State consider giving potassium iodide to the general public as a protective measure, supplementing the evacuation and sheltering protective actions.   The NRC found that potassium iodide is a reasonable, prudent, and inexpensive supplement to evacuation and sheltering for specific local conditions.   For States that choose to give potassium iodide to the general public as part of their emergency plans, the NRC funded an initial supply and replenishment of expired potassium iodide tablets.   To date, 23 States have asked for and received potassium iodide tablets, which the NRC distributes in 65 milligram pills.   In January 2002, the NRC, in cooperation with the cognizant agencies, updated the Federal policy statement on potassium iodide prophylaxis to reflect the changes in NRC regulations.   In September 2006, the Commission approved replenishment plans for initial State supplies.

16.5   Inspection Practices—Reactor Oversight Process for Emergency Preparedness

The NRC's Reactor Oversight Process addresses emergency preparedness.   The process allows licensees to manage their own emergency preparedness programs, including corrective actions, as long as the performance indicators and inspection findings are within an acceptable performance band.   The NRC handles inspection findings through its significance determination process.   Article 6 of this report discusses the NRC's Reactor Oversight Process and significance determination process.

Emergency preparedness is a component of the Reactor Oversight Process, one of its seven cornerstones of safety.   The objective of this cornerstone is to "ensure that the licensee is capable of implementing adequate measures to protect the public health and safety during a radiological emergency."   Oversight of this cornerstone is achieved through three performance indicators and a supporting risk-informed inspection program.   The performance indicators are drill and exercise performance, emergency response organization drill participation, and alert and notification system reliability.   The performance indicator for drill and exercise performance monitors timely and accurate licensee performance in drills, exercises, and actual events when presented with opportunities to classify emergencies, notify offsite authorities, and recommend protective actions.   The indicator for emergency response organization drill participation measures the percentage of key members of the licensee's emergency response organization who have participated in proficiency-enhancing drills, exercises, training opportunities, or an actual event over a determinant amount of time.   The alert and notification system reliability indicator monitors the reliability of the offsite alert and notification system, which is a critical link for communicating with the public.

The emergency preparedness cornerstone of the Reactor Oversight Process includes the following areas for inspection:

- Correction of Emergency Preparedness Weaknesses - Inspectors evaluate the licensees' programs on problem identification and resolution for emergency preparedness.

- Drill Evaluation - Inspectors evaluate drills and simulator-based training evolutions in which shift operating crews and licensee emergency response organization members participate.

- Exercise Evaluation - Inspectors independently observe the licensee's performance in classifying, notifying, and developing recommendations for protective actions, and other activities during the exercise. The inspectors also ensure that the licensee's self-critique is consistent with their observations.

- Alert and Notification System Evaluation - Inspectors verify how well the testing program complies with program procedures.

- Emergency Action Level Changes - Inspectors review all of the licensee's changes to emergency action levels to determine if any of the changes have decreased the effectiveness of the emergency plan.

- Emergency Response Organization Staffing and Augmentation System - Inspectors review the augmentation system to determine whether, as designed, it will support augmentation of the emergency response organization in accordance with the goals for activating the emergency response facility.

- Reactor Safety/Emergency Preparedness - Inspectors verify that the data reported for the performance indicator values are valid.

- Emergency Plan Changes - Inspectors sample changes to the emergency plan to ensure that the effectiveness of the emergency plan has not decreased.

- Force-on-Force Exercise Evaluation - Inspectors primarily assess the nuclear plant's physical protection strategy to defend against the design basis threat. A full inspection, spanning several weeks, includes both table-top drills and exercises, which simulate combat between a mock commando-type adversary force and the nuclear plant security force. As part of these inspections, the NRC's inspectors assess the licensee's integration of emergency response actions into its overall response to the threats.

It is important to note, however, that even though FEMA has no direct regulatory authority over State or local governments and their full-participation exercise evaluations are not considered inspections, FEMA's exercise findings carry substantial weight in the NRC regulatory process. FEMA notifies the State government and the NRC of any significant deficiencies in offsite performance shortly after the exercise. FEMA also issues a formal exercise report within 90 days of the exercise's completion describing the FEMA exercise findings. Because of the potential effect of deficiencies on offsite emergency preparedness, findings are expected to be corrected within 120 days of the exercise. Failure of offsite organizations to correct deficiencies promptly could lead FEMA to withdraw its finding of "reasonable assurance." This would cause the NRC to assess the continued operation of the facility.

16.6 Responding to an Emergency

Fundamental changes in the response to national emergencies have occurred as a result of the publication of the National Response Framework in January 2008 and the update of its associated annexes. Additionally, DHS has revised and republished the National Incident Management System (NIMS) document in December 2008.

This section explains the roles of the NRC, other Federal agencies, licensees, States, and local governments during the response to an incident. It also explains the security issues associated with supporting the response efforts.

16.6.1 Federal Response

The Federal response structure has been revamped with the creation of DHS and the implementation of Homeland Security Presidential Directive 5, "Management of Domestic Incidents," dated March 4, 2003. This directive establishes the Secretary of Homeland Security as the primary Federal official for managing domestic incidents. Under the Homeland Security Act of 2002, DHS is responsible for coordinating Federal operations within the United States to prepare for, respond to, and recover from terrorist attacks, major disasters, and other emergencies.

DHS will assume overall Federal incident management coordination responsibilities when any one of the following four conditions applies:

(1)   A Federal department or agency acting under its own authority has requested DHS assistance.

(2)   The resources of State and local authorities are overwhelmed, and the appropriate State and local authorities have requested Federal assistance.

(3)   More than one Federal department or agency has become substantially involved in responding to the incident.

(4)   The President of the U.S. has directed the Secretary to assume incident management responsibilities.

In 2008, the governing documents outlining the responsibilities of the Secretary of Homeland Security, DHS, and other Federal, State, and local entities were updated. These documents were related to NIMS and the National Response Framework and its associated annexes.

NIMS is a comprehensive, national approach to incident management that is applicable at all jurisdictional levels and across functional disciplines. NIMS enables Federal, State, and local entities to work together to prevent, protect against, respond to, recover from, and mitigate the effects of incidents, regardless of cause, size, location, or complexity, in order to reduce the loss of life and property and harm to the environment. NIMS provides an organized set of scalable and standardized operational structures that is critical for allowing various organizations and agencies to work together in a predictable, coordinated manner.

NIMS works hand-in-hand with the National Response Framework. NIMS provides the template for the management of incidents, while the National Response Framework describes the structures and mechanisms for national-level policy for incident management. The National

Response Framework provides guidance on Federal coordinating structures and processes to prepare for, respond to, and recover from domestic incidents such as terrorist attacks, major disasters, and other emergencies.

The Federal response to a potential nuclear or radiological incident is designed to support the efforts of the facility operator and offsite officials. For such emergencies, Federal response activities are carried out in accordance with the National Response Framework's Nuclear/Radiological Incident Annex, which describes the roles of DHS, coordinating agencies (e.g., the NRC during an incident with one of its licensees), and other supporting Federal agencies. During an incident that meets the criteria of Homeland Security Presidential Directive 5 (invoked during a terrorist-related incident or at a general emergency level for an NRC licensee), DHS is responsible for the overall domestic incident management, while the coordinating agency coordinates the Federal on-scene actions and helps State and local governments determine measures to protect life, property, and the environment. The coordinating agency may respond as part of the Federal response as requested by DHS under the framework, or in accordance with its own authorities. During less severe incidents, coordinating agencies will oversee the onsite response, monitor and support owner or operator activities (when there is an owner or operator), provide technical support to the owner or operator if asked, serve as the principal Federal source of information about onsite conditions, and, if asked, advise the State and local government agencies on implementing protective actions. The coordinating agency will also provide a hazard assessment of onsite conditions that might have significant offsite effects and ensure that onsite measures are taken to mitigate offsite consequences.

16.6.2   Licensee, State, and Local Response

The NRC recognizes the nuclear power plant operator (licensee) and the State or local government as the two primary decisionmakers during a radiological incident at a licensed power reactor. The licensee is primarily responsible for mitigating the consequences of an incident on site and recommending timely protective actions to State and local authorities. The States or local governments are ultimately responsible for implementing appropriate protective actions for public health and safety.

16.6.3   The NRC's Response

In fulfilling its legislative mandate to protect the public health and safety, the NRC has developed a plan and procedures detailing its response to incidents involving licensed material and activities (NUREG-0728, Revision 4, "NRC Incident Response Plan," dated April 14, 2005). In accordance with that plan, the NRC will initially assess any reported event and decide whether or how it will respond as an agency. To meet its statutory and regulatory obligations as the coordinating agency, the NRC will usually dispatch a team to the site for all serious incidents. The team may help the State interpret and analyze technical information, update other responding Federal agencies on event conditions, and coordinate any multiagency Federal response.

Once the NRC has decided to respond as an agency, it activates the NRC Headquarters Operations Center in Washington, DC, and the associated regional incident response center. The NRC Headquarters Operations Center will then take the following actions: (1) maintain continuous communications with the facility, (2) assess the incident, (3) advise the facility operator and offsite officials, (4) coordinate the Federal radiological response with other Federal agencies, and (5) respond to inquiries from the national media. The staff at the NRC

Headquarters Operations Center includes emergency preparedness and response experts and personnel experienced with liaison activities. Because regional office personnel usually have firsthand knowledge of the details of the affected facility, early in an incident the Regional Administrator provides operational authority from the affected regional office and, if necessary, from the regional incident response center. When NRC onsite presence is required, the NRC will dispatch a team from the affected regional office.

As soon as the NRC site team arrives at the facility and is ready to assume the agency's leadership role, it may be delegated certain responsibilities that may include the authority to direct the agency's on site response.

The NRC site team consists of many technical specialists and representatives who respond to the designated response centers used by the facility and offsite officials to coordinate the response. These response centers include the affected State's emergency operations center, the first-responder's incident command post, the joint information center, established by the facility or local government to interact with the media, and, if necessary, the joint field office (the primary Federal incident management field structure that is usually established 48 to 72 hours after an incident). Through participation in these response centers, the NRC site team has access to wide-ranging State and Federal response assets, as well as to extensive radiological monitoring capabilities through DOE (i.e., field teams and aerial monitoring).

The NRC regularly participates in nuclear power plant and Federal interagency exercises each year to ensure its readiness to respond. The NRC also participates in the planning and conduct of the Eagle Horizon and National Level exercises each year. The NRC's participation in such exercises gives the agency a valuable perspective on multi-event response. This perspective improves interagency cooperation and imparts a better understanding of response roles during emergencies.

16.6.4   Aspects of Security that Support Response

Before September 11, 2001, the security measures at nuclear facilities provided reasonable assurance that public health and safety would be protected in the event of an attack encompassed by the design-basis threats for radiological theft and sabotage, which are described in 10 CFR 73.1, "Purpose and Scope." Since September 11, 2001, the nuclear industry has significantly enhanced its defensive capability through the voluntary actions taken by licensees in response to NRC advisories and as required by the orders issued on February 25, 2002, January 7, 2003, and April 29, 2003. The enhancements outlined in the orders include security measures against threats from an insider, waterborne attack, vehicle bomb attack, and land-based assault. The three orders issued on April 29, 2003, also identified a revised design basis threat against which licensees must be prepared to defend. The NRC has codified through rulemaking many of the security requirements that it newly imposed on licensees by order following September 11, 2001. The NRC will consider additional measures in the future as necessary. (The Other Major Regulatory Accomplishments section of this report provides more details about the power reactor security rulemaking.)

The NRC receives a substantial and steady flow of information from the national intelligence community, law enforcement, and licensees and continually evaluates this information to assess threats to regulated facilities or activities. The NRC works with a variety of other Federal agencies, particularly DHS and the Federal Bureau of Investigation, to ensure that security around nuclear power plants is well coordinated and that law enforcement responders are prepared for a significant event. If an event were to occur, the NRC would have significant resources accessible to it and as many as 18 Federal agencies available to help mitigate the radiological consequences of a serious accident or successful attack.

16.7   International Arrangements

The NRC has agreements with its neighbors, principally Canada and Mexico, and commitments to IAEA.

Under its signed agreements with Canada and Mexico, the NRC will promptly notify and exchange information in the event of an emergency that has the potential for trans-boundary effects.   The agreement with Canada, "Agreement Between the Government of the United States of America and the Government of Canada on Cooperation in Comprehensive Civil Emergency Planning and Management," is implemented by the procedure specified in "Administrative Arrangement Between the United States Nuclear Regulatory Commission and the Atomic Energy Control Board of Canada for Cooperation and the Exchange of Information in Nuclear Regulatory Matters," both dated June 21, 1989.   The agreement between the NRC and the Canadian Nuclear Safety Commission, which replaced the Atomic Energy Control Board, was most recently renewed in 2007.

The agreement with Mexico, "Agreement for the Exchange of Information and Cooperation in Nuclear Safety Matters," is implemented by the "Implementing Procedure for the Exchange of Technical Information and Cooperation in Nuclear Safety Matters Between the Nuclear Regulatory Commission of the United States of America and the Comision Nacional de Seguridad Nuclear y Salvaguardias of Mexico," both dated October 6, 1989.   This agreement was most recently renewed in 2007.

To meet the U.S. commitment under the IAEA Convention on Early Notification of a Nuclear Accident, the NRC will promptly notify IAEA if a serious accident occurs at a commercial nuclear power plant.   Afterward, the NRC will work with the U.S. Department of State to update IAEA.

Since 2001, the United States has fully participated in the International Nuclear Event Scale by evaluating operating reactor events and reporting to IAEA any events resulting in a categorization of International Nuclear Event Scale Level 2 or higher.

ARTICLE 17.   SITING

Each Contracting Party shall take the appropriate steps to ensure that appropriate procedures are established and implemented for

(i)   evaluating all relevant site-related factors that are likely to affect the safety of a nuclear installation for its projected lifetime

(ii)   evaluating the likely safety impact of a proposed nuclear installation on individuals, society, and the environment

(iii)   re-evaluating, as necessary, all relevant factors referred to in subparagraphs (i) and (ii) so as to ensure the continued safety acceptability of the nuclear installation

(iv)   consulting Contracting Parties in the vicinity of a proposed nuclear installation, insofar as they are likely to be affected by that installation and, upon request, providing the necessary information to such Contracting Parties, in order to enable them to evaluate and make their own assessment of the likely safety impact on their own territory of the nuclear installation

This section explains the NRC's responsibilities for siting, which include site safety, environmental protection, and emergency preparedness.   First, this section discusses the regulations applying to site safety and their implementation, emphasizing regulations applying to seismic, geological, hydrological, meteorological, and radiological assessments.   Next, it explains environmental protection.   Article 16 of this report discusses emergency preparedness and international arrangements, which would apply to Contracting Parties in obligation iv, above.

17.1   Background

The NRC's siting responsibilities stem from the Atomic Energy Act, the Energy Reorganization Act, and the National Environmental Policy Act.   These statutes confer broad regulatory powers on the Commission and authorize the NRC to promulgate regulations that it deems necessary to fulfill its responsibilities under the acts.

The NRC's siting regulations are integral to protecting public health and safety and the environment.   Siting away from densely populated centers has been, and will continue to be, an essential component of the NRC's defense-in-depth safety philosophy (see Article 18 of this report), which also includes multiple-barrier containment and redundant and diverse safety systems.   The primary factors that determine public health and safety are reactor design and construction and operation of the facility.   However, siting factors and criteria are important to ensure that radiological doses from normal operation and postulated accidents will be acceptably low, natural phenomena and man-made hazards will be properly accounted for in the design of the plant, and the human environment will be protected during the construction and operation of the plant.

For the first time since the 1970s, the nuclear power industry in the United States is seeking approval for sites that could host new nuclear power plants.   To ensure that the agency can effectively carry out its responsibilities associated with, among others, an early site permit application, the NRC consolidated regulatory functions to (1) manage near-term future licensing activities, (2) work with stakeholders on new reactor licensing activities, and (3) assess the NRC's readiness to perform new reactor licensing reviews.

In 2003, applicants submitted three early site permit applications to the NRC for sites in Virginia, Illinois, and Mississippi; in 2007, the NRC issued the three early site permits. In 2006, an applicant submitted an early site permit application for a site in Georgia with a subsequent request for authorization to perform limited work; in 2009 the NRC issued the permit and the limited work authorization. These four sites are near existing nuclear power plants, which enables the applicants to use existing physical and administrative infrastructures, programs, and siting information and to reduce the effects on the environment compared with using an undeveloped location. In 2010, one applicant submitted an early site permit application for a previously undeveloped ("green-field") site in Texas and another applicant submitted an early site permit application for a site near existing nuclear power plants in New Jersey.

In anticipation of these applications and to ensure that future license applicants and the public understand the NRC's process for reviewing programs and siting information, the NRC documented its review process and criteria in RS-002, "Processing Applications for Early Site Permits," dated May 3, 2004.

The NRC received an unprecedented number of applications that require siting evaluations under the combined license application provisions of 10 CFR Part 52. While many of these applications were for locations close to existing facilities, some will be at locations where applicants requested construction permits under 10 CFR Part 50 but plants were not completed, and others at "green-field" sites. In 2007, applicants submitted five combined license applications for a total of eight units for sites in Texas, Alabama, Maryland, Virginia, and South Carolina. In 2008, applicants submitted 11 combined license applications for a total of 16 units for sites in North Carolina, Mississippi, South Carolina, Florida, Michigan, Texas, Louisiana, Missouri, New York, and Pennsylvania. In 2009, one applicant submitted a combined license application for two units at a site in Florida.

17.2 Safety Elements of Siting

This section explains the safety elements of siting. After providing a short background, it explains seismic and geological assessments. Finally, it discusses radiological assessments performed for initial licensing, as a result of facility changes, and according to regulatory developments since the licensing of all U.S. operating plants.

17.2.1 Background

The NRC's site safety regulations consider societal and demographic factors, manmade hazards (such as airports and dams), and physical characteristics of the site (such as hydrological, seismic, and meteorological factors) that could affect the design of the plant. The requirements are specified in 10 CFR Part 100, "Reactor Site Criteria," Appendix A, "Seismic and Geologic Siting Criteria for Nuclear Power Plants," 10 CFR Part 100, Subpart B, "Evaluation Factors for Stationary Power Reactor Site Applications on or after January 10, 1997," and 10 CFR 100.23, "Geologic and Seismic Siting Criteria." The requirements in 10 CFR 100.23 apply to applicants for an early site permit, a combined license, a construction permit, or an operating license on or after January 10, 1997. RGs 1.27, Revision 2, "Ultimate Heat Sink for Nuclear Power Plants," dated January 1976; RG 1.59, Revision 2, "Design Basis Floods for Nuclear Power Plants," dated August 1977; RG 1.102, Revision 1, "Flood Protection for Nuclear Power Plants," dated September 1976; and RG 1.208, "A Performance-Based Approach to Define the Site-Specific

Earthquake Ground Motion," dated March 2007, describe methods acceptable to NRC staff for implementing those requirements.

The applicant's safety analysis report must describe the physical characteristics in and around the site and contain accident analyses that are relevant to evaluating the suitability of a site. A number of RGs provide guidance on issues of site safety that applicants need to address. NUREG-0800 guides the staff in reviewing the site safety content of these reports. RS-002 identifies parts of NUREG-0800 that apply to the review of early site permits.

Once licensed to operate, the licensee is expected to monitor the environs around the nuclear power plant and report in its safety analysis report changes in the environs that may affect the continued safe operation of the facility.

17.2.2 Assessments of Seismic and Geological Aspects of Siting

The NRC's siting regulations listed in Section 17.2.1 of this report detail the assessments applying to seismic and geologic aspects of siting. Recent developments in assessments include the performance-based approach for determining the site-specific ground motion response spectrum and the safe-shutdown earthquake. The performance-based approach combines the site seismic hazard curves and seismic fragility curves for nuclear structures to meet a specified performance target. RG 1.208, which the NRC developed as a replacement for RG 1.165, "Identification and Characterization of Seismic Sources and Determination of Safe Shutdown Earthquake Ground Motion," dated March 1977, describes this new approach in detail.

RG 1.208 also incorporates recent developments in seismic hazard assessment, including the use of cumulative absolute velocity filtering in place of a lower-bound magnitude cutoff and guidance on the development of earthquake time histories, site response analysis, and the location of the ground motion response spectrum within the soil profile.

In 2003, the three early site permit applicants used the EPRI central and eastern U.S. seismic source models as a starting point for their site applications. Applicants updated the EPRI source models to reflect advances in central and eastern U.S. seismic and geologic source modeling. In 2004, EPRI also updated its ground motion models for generic use in new plant probabilistic seismic hazard analyses for sites located in the central and eastern U.S.

The NRC reviews and certifies advanced reactor designs under 10 CFR Part 52. The designs use high seismic design input that is independent of any site but capable of being sited in most currently existing sites. The NRC requires all new and advanced reactor designs to demonstrate that they have a plant-level seismic margin of 1.67 times the design-basis safe-shutdown earthquake with high confidence (i.e., 95 percent) in low (i.e., 5 percent) probability of failure.

17.2.3 Assessments of Radiological Consequences

The Reactor Site Criteria Rule, 10 CFR Part 100, is the regulation under which all U.S. operating plants were licensed. It contains provisions for assessing whether radiological doses from postulated accidents will be acceptably low. The NRC has issued the following regulatory guidance for licensees to implement the requirements regarding of 10 CFR Part 100:

- RG 1.3, Revision 2, "Assumptions Used for Evaluating the Potential Radiological

Consequences of a Loss-of-Coolant Accident for Boiling-Water Reactors," dated June 1974

- RG 1.4, Revision 2, "Assumptions Used for Evaluating the Potential Radiological Consequences of a Loss-of-Coolant Accident for Pressurized-Water Reactors," dated June 1974

- RG 1.145, Revision 1, "Atmospheric Dispersion Models for Potential Accident Consequence Assessments at Nuclear Power Plants," dated November 1982

Although applicants analyze dose primarily to support reactor siting, licensees are required to evaluate the potential increase in the consequences of accidents that might result from modifying facility SSCs. Commitments (including the radiological acceptance criteria) made by the applicant during siting and documented in its final safety analysis report remain binding until modified. A licensee must evaluate the potential consequences of design changes against these radiological criteria to demonstrate that the changes will result in a design that still conforms to the regulations and commitments. If the consequences increase more than minimally, as outlined in 10 CFR 50.59 or require a change to the technical specifications, as discussed in Article 14 of this report, the licensee must obtain NRC approval before implementing the proposed modification.

Regulatory developments since the licensing of all U.S. plants now operating include a revision to 10 CFR Part 100 in 1996; NUREG-1465, "Accident Source Terms for Light-Water Nuclear Power Plants," dated February 1995; RG 1.183, "Alternative Radiological Source Terms for Evaluating Design Basis Accidents at Nuclear Power Reactors," dated July 2000, which guided the use of NUREG-1465; and 10 CFR 50.67, "Accident Source Term," which allowed licensees to use alternative source terms.

The NRC has applied the 1996 revision to 10 CFR Part 100, along with the alternative source term, in its design certification review for a passive advanced light-water reactor, the AP600. More recently, the agency has applied the practice to the AP1000 reactor with similar results and is applying it for all contemplated light-water reactor design certification application reviews, including the Economic Simplified Boiling-Water Reactor (ESBWR), the U.S. EPR, and the U.S. Advanced Pressurized Water Reactor (US-APWR). For other than light-water reactor designs, including advanced reactors, applicants will have to describe their rationale for an appropriate accident source term characterization that will be subject to NRC independent review.

The industry continues to explore the use of the alternative source term in implementing cost-beneficial licensing actions at operating reactors. Some of these applications resulted in improved safety equipment reliability and reduced occupational exposures. Since the issuing of 10 CFR 50.67, more than half of the operating reactor licensees requested either full implementation of the alternative source term or selective implementation for certain regulatory applications. Operating plant licensees have also used the alternative source term to analyze the adequacy of certain engineered safety features in meeting the operability requirements in their operating reactor technical specifications.

17.3 Environmental Protection Elements of Siting

This section explains the environmental protection elements of siting. It covers the governing documents and site approval process. Since the last operating plants in the United States received licenses, issues have arisen that must be considered in siting reviews. This section explains the effect of these issues.

17.3.1   Governing Documents and Process

The environmental protection elements of siting consist of the plant's demands on the environment (e.g., water use and effects of construction and operation). These elements are addressed in 10 CFR Part 51, which implements the National Environmental Policy Act consistent with the NRC's statutory authority and reflects the agency's policy to voluntarily apply the regulations of the President's Council on Environmental Quality, subject to certain conditions. Integrating environmental reviews into its routine decisionmaking, the NRC considers environmental protection issues and alternatives before taking any action that may significantly affect the human environment.

The site approval process leading to the construction or operation of a nuclear power plant requires the NRC to prepare an environmental impact statement. The updated and revised environmental standard review plans (NUREG-1555, "Standard Review Plans for Environmental Reviews for Nuclear Power Plants," dated March 2000) guide the staff's environmental reviews for a range of applications, including green field site reviews for construction permits and operating licenses under 10 CFR Part 50, for early site permits under 10 CFR Part 52, Subpart A, "Early Site Permits," and for combined licenses under 10 CFR Part 52, Subpart C, "Combined Licenses," when the application does not reference an early site permit. The NRC issued updates to review practices in 2007 and 2010 to reflect experience gained from early site permit reviews, account for the changes resulting from the amendment to the limited work authorization rule (discussed later in this section), and include consideration of the environmental effects of greenhouse gas emissions and climate change. Article 19 of this report, in RG 1.206, "Combined Operating Licenses for Nuclear Power Plants," dated June 2007, and RS-002, dealing with early site permits, discuss these governing documents and processes. Environmental standard review plans are also appropriate for environmental reviews of applications for combined licenses under 10 CFR Part 52, Subpart C, when the applications reference an early site permit. Reviews of early site permit applications are limited because the reviews focus on the environmental effects of reactor construction and operation that have characteristics that fall within the postulated site parameters and because the reviews need not assess benefits (e.g., the need for power) or alternative energy sources. The environmental information in applications for combined licenses that reference an early site permit is limited to (1) information to demonstrate that the design of the facility falls within the parameters specified in the early site permit, (2) new and significant information on issues previously considered in the early site permit proceeding, and (3) any significant environmental issue not considered in any previous proceeding on the site or design.

The environmental standard review plans in Supplement 1 to NUREG-1555 guide the staff's environmental review for license renewal applications under 10 CFR Part 54. Article 14 of this report discusses the license renewal process in more detail.

Several other NRC actions on siting and site suitability require environmental reviews, including issuance of limited work authorizations (10 CFR 50.10(e); 10 CFR 52.25, "Extent of Activities Permitted"; and 10 CFR 52.91, "Authorization to Conduct Site Activities"), early partial decisions (10 CFR 2.600, "Scope of Subpart," in Subpart F, "Additional Procedures Applicable to Early Partial Decisions on Site Suitability Issues in Connection with an Application for a Permit to Construct Certain Utilization Facilities," of 10 CFR Part 2), and pre-application early reviews of site suitability issues (Appendix Q, "Pre-application Early Review of Site Suitability Issues," to

10 CFR Part 50).

With its 2007 amendment to the limited work authorization licensing framework (10 CFR 50.10, "License Required; Limited Work Authorization"), the Commission limited its authority to construction activities that have a "reasonable nexus to radiological health and safety or common defense and security" and defined "construction" within the context of its authority. The effect of this change is not limited to limited work authorizations. Other activities related to building the plant that do not require NRC approval (but may require a permit from other regulatory agencies) may occur before, during, or after NRC-authorized construction activities. These activities called "preconstruction" in 10 CFR 51.45(c), may be regulated by other local, State, Tribal or Federal agencies. On September 12, 2008, the NRC and the U.S. Army Corps of Engineers signed an updated memorandum of understanding to enhance the effectiveness of reviews of nuclear power plant license applications that would require multiple Federal permits under separate statutes. The NRC and the U.S. Army Corps of Engineers are participating as cooperating agencies in the preparation of many environmental impact statements.

17.3.2   Other Considerations for Siting Reviews

Since the NRC last issued construction permits under 10 CFR Part 50 in the 1970s and coincident with the publication of the initial environmental standard review plan, many changes to the regulatory environment have affected the NRC and applicants seeking site approvals. These include new environmental laws and regulations, changes in policies and procedures resulting from decisions of courts and administrative hearing boards, and changes in the types of authorizations, permits, and licenses issued by the NRC. This section highlights some of these changes and their effects on the environmental standard review plans.

In the late 1980s, the NRC issued regulations that gave an alternative licensing framework to 10 CFR Part 50, which required a construction permit followed by an operating license. The new framework in 10 CFR Part 52 introduced the concepts of approving designs independent of sites and approving sites independent of designs, and then efficiently linked the approvals to approve construction and operate the facility. As discussed in Section 17.1 of this report, the NRC has received four early site permit applications under 10 CFR Part 52 and is actively conducting siting reviews.

Toward that end, the NRC issued RS-002, which embodies the environmental guidance in NUREG-1555, the environmental standard review plan, and the outcome of interactions with stakeholders. In addition, in 2007, the NRC revised 10 CFR Part 52 to reflect experience gained in its use and to provide guidance on the preparation of combined license applications; as part of that rulemaking the NRC issued RG 1.206, which includes guidance on the assessment of environmental issues.

As described in previous U.S. National Reports, other relevant regulatory developments include the following:

- Presidential Executive Order 12898, "Federal Actions to Address Environmental Justice in Minority and Low-Income Populations," dated February 1994, which instructed Federal agencies to make "environmental justice" part of each agency's mission by addressing disproportionately high and adverse human health or environmental effects of Federal programs, policies, and activities on minority and low-income populations

- the Yellow Creek Decision, which determined that the authority of the NRC is limited in matters that are expressly assigned to EPA

- changes in the economic regulation of utilities that have expanded the options to be addressed in considering the need for power in environmental impact statements

- design alternatives to mitigate the consequences of severe accidents

- EPA rules about cooling water intake structures

17.4 Consultation with Other Contracting Parties To Be Affected by the Installation

At this time, the NRC does not have any specific international arrangements with neighboring states for siting new builds. However, the agency's current arrangements with its Canadian and Mexican regulatory counterparts for the exchange of information and experience would serve as the mechanism for any cooperative dialogue if such a situation arose.

ARTICLE 18. DESIGN AND CONSTRUCTION

Each Contracting Party shall take the appropriate steps to ensure that:

(i)   **the design and construction of a nuclear installation provides for several reliable levels and methods of protection (defense in depth) against the release of radioactive materials, with a view to preventing the occurrence of accidents and to mitigating their radiological consequences should they occur**

(ii)   **the technologies incorporated in the design and construction of a nuclear installation are proven by experience or qualified by testing or analysis**

(iii)   **the design of a nuclear installation allows for reliable, stable, and easily manageable operation, with specific consideration of human factors and the man-machine interface**

This section explains the defense-in-depth philosophy and how it is embodied in the general design criteria of U.S. regulations.   It explains how applicants meet the defense-in-depth goals and how the NRC reviews applications and conducts inspections before issuing licenses to ensure that this philosophy is implemented in practice.   Next, this section discusses measures for ensuring that the applications of technologies are proven by experience or qualified by testing or analysis.   Finally, this section discusses requirements for reliable, stable, and easily manageable operation, specifically considering human factors and the man-machine interface. Article 12 of this report also provided information on these obligations.

18.1 Defense-in-Depth Philosophy

This section explains the defense-in-depth philosophy followed in regulatory practice, governing documents and regulatory process for designing and constructing a nuclear power plant.   It also discusses relevant experience and examples.

18.1.1 Governing Documents and Process

The defense-in-depth philosophy, as applied in regulatory practice, requires that nuclear plants contain a series of independent, redundant, and diverse safety systems.   The physical barriers for defense in depth in a light-water reactor are the fuel matrix, the fuel rod cladding, the primary coolant pressure boundary, and the containment.   The levels of protection in defense in depth are (1) a conservative design, quality assurance, and safety culture, (2) control of abnormal operation and detection of failures, (3) safety and protection systems, (4) accident management, including containment protection, and (5) emergency preparedness.

Appendix A to 10 CFR Part 50 embodies the defense-in-depth philosophy.   General design criteria cover protection by multiple fission product barriers, protection and reactivity control systems, fluid systems, containment design, and fuel and radioactivity control.   The NRC staff amplified its defense-in-depth philosophy in RG 1.174, which provides guidance on using a PRA in risk-informed decisions on plant-specific changes.   The general design criteria establish the minimum requirements for the principal design criteria, which in turn establish the necessary design, fabrication, construction, testing, and performance requirements for SSCs that are important to safety.

To ensure that a plant is properly designed and built as designed, that proper materials are used in construction, that future design modifications are controlled, and that appropriate maintenance and operational practices are followed, a good quality assurance program is needed. To meet this need, 10 CFR Part 50, Appendix A, General Design Criterion 1, "Quality Standards and Records," and its implementing regulatory requirements specified in 10 CFR Part 50, Appendix B, establish quality assurance requirements for all activities affecting the safety-related functions of the SSCs.

Pursuant to the two-step licensing process under 10 CFR Part 50, an applicant for a construction permit must present the principal design criteria for a proposed facility in its preliminary safety analysis report (see 10 CFR 50.34, "Contents of Applications; Technical Information"). For guidance in writing a safety analysis report, the applicant may use RG 1.70. The safety analysis report must also contain design information for the proposed reactor and comprehensive data on the proposed site. The report must also discuss various hypothetical accident situations and the safety features to prevent accidents or, if accidents occur, to mitigate their effects on both the public and the facility's employees.

After obtaining a construction permit under 10 CFR Part 50, the applicant must submit a final safety analysis report to support an application for an operating license, unless it submitted the report with the original application. This report should give the details of the final design of the facility, plans for operation, and procedures for coping with emergencies. The preliminary and final safety analysis reports are the principal documents the applicant provides for the staff to determine whether the proposed plant can be built and operated without undue risk to the health and safety of the public. The NRC expects that future applications to build nuclear power plants will use the combined license process under 10 CFR Part 52. Applications submitted under 10 CFR Part 52 must meet all of the 10 CFR Part 50 requirements. A significant difference in the 10 CFR Part 52 process is that the final safety analysis report must be submitted before authorization is granted to begin construction. Article 19 of this report describes the combined license review process.

The NRC staff reviews safety analysis reports according to NUREG-0800 to ensure that the applicant has satisfied the general design criteria and other applicable regulations. The staff reviews each application to determine whether the plant design meets the Commission's regulations (10 CFR Part 20, 10 CFR Part 50, 10 CFR Part 73, "Physical Protection of Plants and Materials," and 10 CFR Part 100). These reviews include, in part, the characteristics of the site. In addition, each application for a nuclear installation must include a comprehensive environmental report that provides a basis for evaluating the environmental impact of the proposed facility. RG 4.2, Revision 2, "Preparation of Environmental Reports for Nuclear Power Stations," dated July 1976, gives applicants information on writing environmental reports. The NRC staff reviews the environmental reports according to NUREG-1555. In reviewing an application, the staff, supported by outside experts, conducts independent technical studies to review certain safety and environmental matters. The staff states its conclusions in an environmental impact statement and a safety evaluation report, which it may update before granting the license. Under the two-step licensing process in 10 CFR Part 50, the NRC does not issue an operating license until construction is complete and the Commission makes the findings required under 10 CFR 50.57, "Issuance of Operating License." For applications submitted under 10 CFR Part 52, the Commission must find that all acceptance criteria in the combined license are met before operation of the facility.

The NRC monitors nuclear power plant construction to ensure compliance with the agency's regulations to protect public health and safety and the environment.   In anticipation that future applicants for construction of a nuclear power plant will apply for a combined license, the NRC has developed an inspection program for future nuclear plants licensed under 10 CFR Part 52.

The new inspection program revises the 10 CFR Part 50 Construction Inspection Program.   It incorporates inspections, tests, analyses, and acceptance criteria (ITAAC) from 10 CFR Part 52, as well as lessons learned from the inspection program used in the previous construction era (1970-1980).   It also considers modular construction at remote locations.

Before construction, the NRC inspection program focuses on the applicant's establishment of a quality assurance program to verify that applications submitted to the NRC meet specified requirements in 10 CFR Part 52 and are of a quality suitable for docketing.   Inspection Manual Chapter 2501, "Construction Inspection Program: Early Site Permit (ESP)," dated October 3, 2007, lists inspections for this phase.

Once the NRC receives an application, the inspection program focuses on supporting the NRC staff's preparation for the mandatory Atomic Safety and Licensing Board hearing and the final Commission decision on whether a combined license should be granted.   Inspection Manual Chapter 2502, "Construction Inspection Program: Pre-Combined License (Pre-COL) Phase," dated October 3, 2007, lists inspections for this phase.

The NRC also interacts with manufacturers and suppliers of safety-related components through the NRC vendor inspection programs that inspect compliance with quality assurance and defect reporting requirements.   Vendor inspections are conducted at vendor shops principally to examine whether the vendor has been complying with 10 CFR Part 50, Appendix B, as required by procurement contracts with applicants and licensees.   Inspection Manual Chapter 2507, "Construction Inspection Program: Vendor Inspections," dated April 27, 2010, lists inspections for vendors.

During construction, inspectors sample the spectrum of the applicant's activities related to the ITAAC in the design-basis document to confirm that the applicant is adhering to quality and program requirements.   NRC inspectors will verify successful ITAAC completion on a sampling basis and will review all ITAAC.   The NRC will publish notices in the *Federal Register* of completed ITAAC.   Additionally, regional specialists inspect and monitor activities at the construction sites.   The NRC will increase the number of resident inspectors stationed in construction sites.   It is expected that the peak resident staffing will be approximately five inspectors at construction sites with one unit and seven at construction sites with two units. Inspection Manual Chapter 2503, "Construction Inspection Program: Inspections of Inspections, Tests, Analyses, and Acceptance Criteria (ITAAC)," dated October 3, 2007, lists inspections for this phase.

As the applicant completes construction, the inspection program focuses on verifying the adequacy of the licensee's preoperational programs such as fire protection, security, training, radiation protection, startup testing, and programs that enable the transition of the organization from construction to power operations.   Inspection Manual Chapter 2504, "Construction Inspection Program—Inspection of Construction and Operational Programs," dated October 15, 2009, lists inspections for this phase.

18.1.2 Experience

18.1.2.1 Regulatory Framework for the Reactivation of Watts Bar Unit 2

The Watts Bar Nuclear Plant, owned by TVA, is located in southeastern Tennessee. The site has two Westinghouse designed PWRs. Watts Bar Unit 1 received a full-power operating license in early 1996 and was the last new power reactor licensed in the U.S. TVA stopped construction at Watts Bar Unit 2 in the mid-1980s. TVA has now resumed Watts Bar Unit 2 construction, and its operating license application is currently pending before the Atomic Safety and Licensing Board. The construction permit for Watts Bar Unit 2 is currently active and expires in 2013.

In its regulatory framework for the completion of Unit 2, TVA proposed and the Commission approved (staff requirements memorandum, dated July 25, 2007, on SECY-07-0096, "Possible Reactivation of Construction and Licensing Activities for the Watts Bar Nuclear Plant Unit 2," dated June 7, 2007) a licensing review approach that employs the current licensing basis for Watts Bar Unit 1 as the reference basis for review and licensing of Unit 2. This approach will ensure safety while preserving design and operational consistency between the units. However, considering the construction status of the unit, the NRC encouraged TVA to adopt updated standards wherever feasible and look for opportunities to resolve any generic safety issues where the unirradiated state of Unit 2 makes the issue easier to resolve before plant operation. The NRC's licensing review will include safety design, environmental review, and inspection of construction activities.

TVA has updated its initial 1970s operating license application. The NRC has published notice of the operating license in the *Federal Register* to provide public notice and an additional opportunity for a hearing. To date, the Southern Alliance for Clean Energy has asked for and received a hearing before an Atomic Safety and Licensing Board. TVA has submitted its final supplemental environmental impact statement for the completion and operation of Watts Bar Unit 2. The NRC has also held public outreach meetings in the vicinity of the site to inform the public about its licensing and inspection activities, including how the public can monitor and participate in the licensing process.

The NRC has established a dedicated team at both at its headquarters and regional offices for review and inspection of the Unit 2 activities. The staff has independently reviewed TVA's regulatory framework and documented its results in a safety evaluation report (NUREG-0847, Supplement 21, "Safety Evaluation Report Related to the Operation of Watts Bar Nuclear Plant, Unit 2," dated February 2009). The review identified the items that must be completed before issuance of an operating license. The NRC Region II office is performing necessary inspections and oversight activities. It developed Inspection Manual Chapter 2517, "Watts Bar Unit 2 Construction Inspection Program," dated February 2008, to provide guidance for these inspection activities. The NRC Region II office is examining historical inspection records, employee concerns, operating experience, scope of new or re-work, and construction deficiency reports. The NRC has established a resident inspector office, with a senior and two resident inspectors dedicated to performing inspections at Watts Bar Unit 2.

As always, safety is the NRC's main focus. Before issuing an operating license, the NRC will confirm that TVA has safely designed and constructed Watts Bar Unit 2 in accordance with regulatory requirements, and that the facility can be safely operated.

The NRC has established a Web page for its Watts Bar Unit activities at
http://www.nrc.gov/reactors/plant-specific-items/watts-bar.html.

18.1.2.2   Design Certifications

For more than 30 years, the Atomic Energy Commission and the NRC have reviewed applications submitted under the two step licensing process in 10 CFR Part 50 and have documented their reviews in safety evaluation reports and supplements for 110 nuclear installations.   Since 1997, the NRC has certified four standard plant designs under the design certification process in 10 CFR Part 52:   GE's advanced BWR (1997), and Westinghouse System 80+ (designed and licensed by Combustion Engineering), AP600, and AP1000 (1997, 2000, and 2006 respectively).   The NRC staff is currently performing the following design certification reviews: GE-Hitachi Nuclear Energy's, ESBWR, Westinghouse's AP1000 design certification amendment, AREVA Nuclear Power's US EPR, Mitsubishi Heavy Industries, Ltd.'s US-APWR and South Texas Project Nuclear Operating Company's Advanced Boiling-Water Reactor (ABWR) design certification application to address the aircraft impact rule.

18.2   Technologies Proven by Experience or Qualified by Testing or Analysis

In 10 CFR 50.43(e), the NRC requires that new technologies are demonstrated to be proven. This rule requires demonstration of new technologies through analysis, appropriate test programs, experience, or a combination thereof.   In its safety analysis reports for the AP600 and AP1000 standard plant designs, Westinghouse used separate effects tests, integral systems tests, and analyses to demonstrate that its passive safety systems will perform as predicted. Section 14.2 of this report discusses the qualification of currently used technologies.

18.3   Design for Reliable, Stable, and Easily Manageable Operation

The NRC specifically considers human factors and the human-system interface in the design of nuclear installations.   For safety analysis reports, the NRC reviews the human factors engineering design of the main control room and the control centers outside of the main control room.   Article 12 of this report also discusses human factors.

18.3.1   Governing Documents and Process

To support its reviews of the human factors engineering issues associated with the certification and licensing of new plant designs, the NRC uses NUREG-0800, Chapter 18, Revision 2, and NUREG-0700, Revision 2, "Human-System Interface Design Review Guidelines," dated May 2002.   The NRC also uses NUREG-0711, Revision 2, "Human Factors Engineering Program Review Model," dated February 2004, for evaluating the design of next-generation main control rooms.   NUREG-0800, Section 14.3.9, "Human Factors Engineering - Inspections, Tests, Analyses, and Acceptance Criteria," dated March 2007, provides additional guidance.   The NRC has recently initiated work to update these review guidelines.   Additionally, the NRC developed guidance for reviewing combined license applications, RG 1.206, which includes sections that address the human factors engineering review of combined license applications.

18.3.2   Experience

The NRC's Office of New Reactors is actively reviewing new plant designs and combined license applications.

18.3.2.1 *Human Factors Engineering*

The NRC is currently conducting design certification reviews of the ESBWR, U.S. EPR, and US-APWR, as well as reviewing applications to amend the design certification rule for the ABWR and AP1000.   The NRC has also received 18 combined license applications that are in various review stages and status.   The NRC's human factors engineering reviews for design certification applications principally focus on evaluating implementation plans for the design of the control facilities to ensure that the design process will be carried out consistent with state-of-the-art human factors principles.   The NRC will verify acceptable implementation of these plans through specified ITAAC (i.e., design acceptance criteria).

18.3.2.2 *Digital Instrumentation and Controls*

Nuclear facility and byproduct licensees are replacing their analog instrumentation and control equipment with digital equipment.   Although digital technology can improve operational performance, the introduction of this technology into nuclear facilities and applications can pose a variety of challenges for the NRC and the nuclear industry:

- the increased complexity of digital technology compared to analog technology
- rapid changes in digital technology that require the NRC to update its knowledge of state-of-the-practice in digital system design, testing, and application
- new failure modes associated with digital technology
- the need to update the acceptance criteria and review procedures used in consistently assessing the safety and security of digital systems

In response to these technical challenges, in January 2007, the NRC formed the Digital Instrumentation and Control Steering Committee.   The Steering Committee focuses on the NRC regulatory activities in progress across several offices, interfaces with the industry on key issues, and facilitates consistent approaches to resolving technical and regulatory challenges.   The members of the Steering Committee include management representatives from the various NRC offices that have regulatory responsibilities related to digital instrumentation and control.

Digital instrumentation and control raises issues that were not relevant to analog systems. Examples of such issues include the following:

- A common-cause failure attributable to software errors was not possible with analog systems. This potential weakness may require consideration of diversity and defense in depth in the application of digital instrumentation and control systems.

- Digital system network architectures raise issues such as interchannel communication, communication between non-safety and safety systems, and cyber security that must be reviewed closely to ensure that public safety is preserved.

- Highly integrated control room designs with safety and nonsafety displays and controls will be the norm for new reactor designs.   Human factors design and quality assurance during all phases of software development, control, and validation and verification are critical.

The Digital Instrumentation and Control Steering Committee has formed seven task working groups focusing on the following key areas of concern:

- cyber security
- diversity and defense in depth
- risk-informed digital instrumentation and control
- highly integrated control room - communications
- highly integrated control room - human factors
- licensing process issues
- fuel cycle facilities

Each of the task working groups developed interim staff guidance for NRC review of new and innovative digital instrumentation and control systems that are found in new reactors and digital upgrades at currently operating reactors.   The guidance also provides the industry with the expectations and criteria that their designs will be evaluated against to determine compliance with NRC regulations.   The staff is using the interim staff guidance in its review of design certifications, combined licenses, and digital upgrades at currently operating reactors.   The staff is in the process of incorporating the interim staff guidance into permanent NRC staff guidance in NUREG-0800 and associated RGs.   The interim staff guidance can be found at http://www.nrc.gov/reading-rm/doc-collections/isg/digital-instrumentation-ctrl.html.

The NRC also actively participates in the Multinational Design Evaluation Program, an international assembly of nuclear regulators addressing common issues with the licensing of new reactors.   The NRC is involved with the Digital Instrumentation and Control Issue-Specific Group, which is looking at ways to harmonize requirements, standards, and guidance for instrumentation and control, and the EPR digital instrumentation and control task group, which is a collaboration of regulators that are reviewing the EPR instrumentation and control design. The Multinational Design Evaluation Program allows the NRC to share digital instrumentation and control information to support regulatory infrastructure improvements and licensing decisions.

18.3.2.3 Cyber Security

After September 11, 2001, the NRC issued two security-related orders, NRC Order EA-02-026, "Issuance of Order for Interim Safeguards and Security Compensatory Measures," dated February 2002, and NRC Order EA-03-086, "Issuance of Order Requiring Compliance with Revised Design Basis Threat for Operating Power Reactors," dated April 2003, that require power reactor licensees to implement measures to enhance cyber security.   These security measures required immediate identification and assessment of computer-based systems deemed to be critical to the operation and security of the facility.   Additionally, licensees were expected to implement any immediate and necessary corrective measures to protect against the cyber threats at the time the orders were issued.

Recognizing that licensees likely used various approaches in the architectural design and implementation of plant computing networks, the NRC began an effort to develop a cyber security self-assessment methodology that could be uniformly applied to U.S.-based nuclear facilities.   Development of such a methodology would provide a means to ensure that the assessments performed by each facility would follow a consistent, repeatable approach, thereby providing comparable metrics to understand the relative cyber security posture of each facility.

The assessment methodology was developed by a multidisciplinary team from Pacific Northwest National Laboratory with input from the NRC and nuclear power industry representatives and issued in October 2004 as NUREG/CR-6847, "Cyber Security Self-Assessment Method for U.S. Nuclear Power Plants." NUREG/CR-6847 provided licensees with information useful for developing an interim cyber security program for their facilities before the codification of cyber security requirements. It does not provide an acceptable means for complying with current cyber security regulations.

Using NUREG/CR-6847 as a foundation, the NEI Cyber Security Task Force developed a comprehensive guidance document, NEI 04-04, "Cyber Security Programs for Power Reactors," dated November 18, 2005, which licensees could use to develop and manage their cyber security programs. In December 2005, the NRC staff accepted NEI 04-04 as an acceptable method for establishing and maintaining a cyber security program at nuclear power plants. At the time of the NRC's endorsement of NEI 04-04, the NRC had not yet proposed comprehensive cyber security regulations.

In March 2009, the NRC issued a new rule on cyber security, 10 CFR 73.54, "Protection of Digital Computer and Communication Systems and Networks." It requires licensees to provide high assurance that nuclear power plants' safety, safety-related, security, and emergency preparedness functions are protected from cyber attacks up to and including the design-basis threat. This new regulation required licensees and combined operating license applicants to submit a cyber security plan, including an implementation schedule, to the NRC for review and approval by November 23, 2009. Essential elements of a plan include describing the process for finding critical digital assets, describing the defensive model (i.e., protective strategy), referencing a comprehensive set of security controls, and describing the process for addressing each control. The cyber security plan must also acknowledge a commitment to maintain the cyber security program and provide adequate documentation of how that will be accomplished.

In January 2010, the NRC published RG 5.71, "Cyber Security Programs for Nuclear Facilities," which provides implementation guidance to licensees and applicants on an acceptable method for satisfying the requirements of 10 CFR 73.54. This guidance describes an acceptable method licensees can follow to address potential security vulnerabilities in each life-cycle phase of critical digital assets that perform safety, safety-related, security, and emergency preparedness functions. It is equally applicable to the combined license applicants and the current fleet of operational reactors. The guidance embodies recommended best practices from standards organizations such as the International Society of Automations, the Institute of Electrical and Electronics Engineers, the National Institute of Standards and Technology, and DHS. In addition, the NRC is in the process of clearly defining the scope of an instrumentation and control review and a cyber security review for the NRC staff and the industry in RG 1.152, Revision 3, "Criteria for Use of Computers in Safety Systems of Nuclear Power Plants," which is currently under review.

In January 2010, the NRC and the North American Electric Reliability Corporation also entered into a 5–year memorandum of understanding to address nuclear plant cyber security roles, responsibilities, and areas of coordination between the two organizations. In essence, the NRC will continue to be responsible for the inspection of digital systems that can affect the safety, security, and emergency preparedness of a nuclear power plant. The North American Electric Reliability Corporation will continue to regulate digital systems related to the generation of electric power. The memorandum of understanding recognizes the need for coordination, information sharing, and incident management and response between the two organizations.

The NRC has implemented a significant and continuing research program in cyber security for digital plant control systems. Also, the NRC is currently in the process of codifying the mandated cyber security enhancement requirements in the two security-related NRC orders by amending its regulations.

18.4   New Reactor Construction Experience Program

The nuclear industry in the United States faced many construction quality and design issues in the 1970s and 1980s.   In 1984, the NRC issued NUREG-1055, "Improving Quality and the Assurance of Quality in the Design and Construction of Nuclear Power Plants," to document the lessons learned from plant construction.   Since then, the NRC has revised some of its licensing review processes and construction oversight programs in order to implement recommendations that were made in NUREG-1055.   In 2007, the NRC began developing a construction experience (ConE) program to focus on collecting, analyzing, and applying lessons learned from the design and construction of new reactors.   To achieve this goal, the NRC staff developed a risk-informed process to obtain, screen, evaluate, communicate, and incorporate construction experience insights into its new reactor licensing and construction oversight activities.

Since 2007, the NRC staff has actively obtained and evaluated ConE information from various domestic and international sources.   The ConE program also reviews all of the operating experience from operating reactors, because the root causes of many events at currently operating reactors date back to the period when these plants were being designed and constructed.   To make the ConE information available and accessible to all NRC staff members, including technical reviewers located at NRC Headquarters and inspectors located in regional offices, the staff has designed and launched a Web-based ConE database.   This database enables all NRC staff to search and retrieve ConE information through word search, plant information, technical discipline, applicable NRC guidance documents, IPs, technical branches, and other methods.   As of February 2010, this database contains about 200 ConE events. Using information in the ConE database, the NRC staff has issued five generic communications in the form of INs to communicate lessons learned from the evaluation of ConE information. The NRC staff continues to actively obtain and evaluate applicable operating and ConE information and plans to develop a publicly available version of its ConE database.   The staff also plans to continue to communicate the lessons learned from the ConE program with the industry and international counterparts through issuing generic communications.

The NRC staff values close cooperation with the international community for the exchange of information on design and construction of new reactors.   The NRC ConE program has been working closely with several countries that are currently building new nuclear power plants. These interactions are carried out through established agency bilateral and multilateral agreements with other countries.   For example, the NRC ConE program staff is contributing to the work of the NEA working group on regulation of new reactors, working group on operating experience, and the European Commission Joint Research Center.   The NRC ConE staff also visits international sites under construction every year to further its cooperation and exchange of technical and regulatory information with other regulatory agencies.   For instance, China's National Nuclear Safety Administration, the French Nuclear Safety Authority, and the Finnish Radiation and Nuclear Safety Authority have recently hosted a number of NRC inspectors at their new reactors construction sites.   These interactions have provided an exceptional hands-on experience for the NRC inspectors to gain a better understanding of the regulatory process and the construction inspection activities in these countries.   Similarly, the NRC has hosted several staff members from foreign nuclear safety regulatory agencies, such as those from China and France, to provide an opportunity for our international counterparts to observe and learn about the licensing process and the oversight of new reactors construction activities in the United States.   The NRC values such partnerships with other regulatory agencies and is committed to continuing its collaborative relationship with the international community to promote nuclear safety, security, and protecting people and the environment.

ARTICLE 19.   OPERATION

Each Contracting Party shall take appropriate steps to ensure that:

(i)   the initial authorization to operate a nuclear installation is based upon an appropriate safety analysis and a commissioning program demonstrating that the installation, as constructed, is consistent with design and safety requirements

(ii)   operational limits and conditions derived from the safety analysis, test, and operational experience are defined and revised as necessary for identifying safe boundaries for operation

(iii)   operation, maintenance, inspection, and testing of a nuclear installation are conducted in accordance with approved procedures

(iv)   procedures are established for responding to anticipated operational occurrences and to accidents

(v)   necessary engineering and technical support in all safety related fields is available throughout the lifetime of a nuclear installation

(vi)   incidents significant to safety are reported in a timely manner by the holder of the relevant license to the regulatory body

(vii)   programs to collect and analyze operating experience are established, the results obtained and the conclusions drawn are acted upon and that existing mechanisms are used to share important experience with international bodies and with other operating organizations and regulatory bodies

(viii)   the generation of radioactive waste resulting from the operation of a nuclear installation is kept to the minimum practicable for the process concerned, both in activity and in volume, and any necessary treatment and storage of spent fuel and waste directly related to the operation and on the same site as that of the nuclear installation take into consideration conditioning and disposal

The NRC relies on regulations in Title 10 of the *Code of Federal Regulations* and internally developed associated programs in granting the initial authorization to operate a nuclear installation and in monitoring its safe operation throughout its life.   This section describes the most significant regulations and programs corresponding to each obligation of Article 19.

19.1   Initial Authorization to Operate

All currently operating reactors in the United States received licenses under the two-step process in 10 CFR Part 50.   This licensing process requires both a construction permit and an operating license.   The additional licensing processes in 10 CFR Part 52 provide for site approvals and design approvals in advance of construction authorization.   In addition, 10 CFR Part 52 includes a process that combines a construction permit and an operating license with conditions into one license (a combined license).   Both the two-step and the combined license processes require NRC approval to construct and operate a nuclear power plant.

The Advisory Committee on Reactor Safeguards, an independent statutory committee established to advise the NRC on reactor safety, reviews each application to construct or operate a nuclear power plant. The committee begins its review early in the licensing process by selecting the proper stages at which to meet with the applicant and NRC staff. Upon completing its review, the committee reports to the Commission.

The public also has an opportunity to have its concerns addressed. The Atomic Energy Act requires that NRC hold a public hearing before it may issue a construction permit, early site permit, or combined license for a nuclear power plant. A three-member Atomic Safety and Licensing Board, consisting of one lawyer who acts as chairperson and two technically qualified persons, conducts the public hearing. Members of the public may submit statements to the licensing board, or they may petition for leave to intervene as full parties in the hearing.

To obtain NRC approval to construct or operate a nuclear power plant, an applicant must submit safety analysis and environmental reports. Article 18 describes the final safety analysis report and the NRC's review of the application for an operating license. A public hearing is neither mandatory nor automatic for an application for an operating license under 10 CFR Part 50. However, soon after the NRC accepts the application for review, it publishes a notice in the *Federal Register* stating that it is considering issuing the license. This notice states that any person whose interest might be affected by the proceeding may petition the NRC for a hearing. If a public hearing is held, the same process applies as for the public hearing for a construction permit.

An early site permit issued under 10 CFR Part 52, Subpart A, provides for resolution of site safety, environmental protection, and emergency preparedness issues, independent of a specific nuclear plant design review. The application for an early site permit must address the safety and environmental characteristics of the site and evaluate potential physical impediments to the development of an acceptable emergency plan or security plan. The applicant may submit additional information on emergency preparedness issues up to a complete emergency plan. The staff documents its findings on site safety characteristics and emergency planning in a safety evaluation report and its findings on environmental protection issues in an environmental impact statement. The early site permit may also allow limited construction activities, subject to redress, before the issuance of a combined license. The NRC will issue a *Federal Register* notice for a mandatory public hearing, and the Advisory Committee on Reactor Safeguards will perform an independent safety review. The duration of an early site permit is 10 – 20 years, and the permit may be renewed. A construction permit or combined license application may reference the early site permit.

The NRC may also certify a standard plant design through a rulemaking under 10 CFR Part 52, Subpart B, "Standard Design Certifications." The design certification process resolves final design information for an essentially complete plant, independent of a specific site, and the Advisory Committee on Reactor Safeguards performs an independent safety review. The NRC has certified four standard plant designs under the design certification process: GE's ABWR, and Westinghouse's System 80+ (originally designed by Combustion Engineering), AP600, and AP1000. The duration of a design certification is 15 years, and the certification may be renewed.

A combined license, issued under 10 CFR Part 52, Subpart C, authorizes construction of a facility in a manner similar to a construction permit under 10 CFR Part 50. An application for a combined license may incorporate by reference an early site permit, design certification, both, or neither. The advantage of referencing an early site permit or design certification is that issues resolved during those processes are not considered at the combined license stage. Just as for a construction permit, the NRC must hold a hearing before the decision to issue a combined license. However, the combined license will specify the inspections, tests, and analyses that the licensee must perform and the acceptance criteria that, if met, are necessary and sufficient to provide reasonable assurance that the facility has been constructed and will be operated in conformity with the license and the applicable regulations.

After issuing a combined license, the NRC staff will verify that the licensee has performed the required inspections, tests, and analyses, and before operation of the facility the Commission must find whether the licensee has met the acceptance criteria. Periodically during construction, the NRC staff will publish notices of the successful completion of inspections, tests, and analyses in the *Federal Register*. Not less than 180 days before the date scheduled for initial loading of fuel, the NRC will publish a notice of intended operation of the facility in the *Federal Register*. An opportunity for a second hearing exists, but petitions for this hearing will be considered only if the petitioner demonstrates that one or more of the acceptance criteria have not been (or will not be) met, and the specific operational consequences of nonconformance would be contrary to providing reasonable assurance of adequate protection of the public health and safety.

19.2   Definition and Revision of Operational Limits and Conditions

The license for each nuclear facility must contain technical specifications that set operational limits and conditions derived from the safety analyses, tests, and operational experience. The regulations contained in 10 CFR 50.36 define the requirements that apply to the plant-specific technical specifications. At a minimum, the technical specifications must describe the specific characteristics of the facility and the conditions for its operation that are required to adequately protect the health and safety of the public. Each applicant must note items that directly apply to maintaining the integrity of the physical barriers that are designed to contain radioactive material. In 10 CFR 50.36 the NRC requires that the technical specifications must be derived from the analyses and evaluations in the safety analysis report. Licensees cannot change the technical specifications without prior NRC approval.

In 1992, the NRC issued improved, vendor-specific (e.g., Babcock and Wilcox, Westinghouse, Combustion Engineering, and GE) standard technical specifications in NUREGs 1430-1434 and periodically revises them on the basis of experience. The NRC issued Revision 3 to these NUREGs in June 2004.

The NRC encourages licensees to use the improved standard technical specifications as the basis for plant-specific technical specifications. The agency also considers requests to adopt parts of the improved standard technical specifications, even if the licensee does not adopt all of the improvements. These parts, which will include all related requirements, will normally be developed as line-item improvements. To date, over half of the operating commercial nuclear plants have converted their technical specifications to the improved standard technical specifications.

Consistent with the Commission's policy statements on technical specifications and the use of PRAs, the NRC and the nuclear industry are developing risk-informed improvements to technical specifications. These improvements and initiatives are intended to maintain or improve safety while reducing unnecessary burden and to make technical specifications congruent with the agency's other risk-informed regulatory requirements (in particular, the risk management requirements of the Maintenance Rule in 10 CFR 50.65(a)(4)).

19.3   Approved Procedures

In the U.S., operations, maintenance, inspection, and testing of a nuclear installation are conducted in accordance with approved procedures.   Each nuclear facility is required to follow the quality assurance requirements in 10 CFR Part 50, Appendix B.   Criterion V "Instructions, Procedures, and Drawings," of Appendix B to 10 CFR Part 50, requires that licensees establish measures to ensure that activities that affect quality will be prescribed by appropriate documented instructions, procedures, or drawings.   RG 1.33, Revision 2, "Quality Assurance Program Requirements (Operation)," dated February 1978, provides supplemental guidance. The NRC addresses the need to perform maintenance according to approved procedures in 10 CFR 50.65, "Requirements for Monitoring the Effectiveness of Maintenance at Nuclear Power Plants."   In 10 CFR 50.65(a)(4) it requires licensees to assess and manage the increase in risk that may result from proposed maintenance activities.

19.4   Procedures for Responding to Anticipated Operational Occurrences and Accidents

The NRC gives recommendations and guidance on procedures for responding to anticipated operational occurrences and accidents in NUREG-0737, "Clarification of TMI Action Plan Requirements," dated November 1980; NUREG-0737, Supplement 1, "Requirements for Emergency Response Capability," dated January 1983; and NUREG-0899, "Guidelines for the Preparation of Emergency Operating Procedures," dated August 1982.

After the 1979 accident at Three Mile Island Unit 2, the NRC issued orders requiring licensees to develop procedures for coping with certain plant transients and postulated accidents.   It also issued NUREG-0737 in 1980 and Supplement 1 to that document in 1983, which recommend that licensees develop procedures to cope with accidents and transients that are caused by initiating events analyzed in the final safety analysis report with multiple failures of equipment.

NUREG-0899 gives programmatic guidance for developing emergency operating procedures. To ensure that proper procedures had been developed to respond to plant transients and accidents, the NRC reviewed each plant using the guidance in NUREG-0800, Section 13.5.2.1.

19.5   Availability of Engineering and Technical Support

The NRC's Reactor Oversight Process, described in Article 6 of this report, includes techniques to ensure that adequate engineering and technical support is available throughout the lifetime of a nuclear installation.   Several of the IPs focus on ensuring the maintenance of adequate support programs.   Licensees also report performance indicators.   Depending on inspection findings and performance indicators, the NRC conducts additional inspections to focus on the causes of the performance problems as prescribed by the Reactor Oversight Process Action Matrix.

19.6   Incident Reporting

Two of the many elements contributing to the safety of nuclear power are emergency response and the feedback of operating experience into plant operations.   The licensee event reporting requirements of 10 CFR 50.72, "Immediate Notification Requirements for Operating Nuclear Power Reactors," and 10 CFR 50.73, "Licensee Event Report System," help to achieve these goals, as 10 CFR 50.72 requires immediate notification requirements via the emergency notification system, and 10 CFR 50.73 requires 60-day written licensee event reports.   All 10 CFR 50.72 event notifications and 10 CFR 50.73 licensee event reports, except those containing sensitive security-related information, are publicly available on the NRC Web site.

The NRC staff uses the information reported under these regulations to respond to emergencies, monitor ongoing events, confirm licensing bases, study potentially generic safety problems, assess trends and patterns of operational experience, monitor performance, identify precursors of more significant events, and provide operational experience to the industry.   Evaluations of events as documented in NRC inspection reports are publicly available on the NRC Web site. The annual abnormal occurrence report to Congress (NUREG-0090), which details specific events that result in a conditional core damage probability greater than 1×10^{-4} and other events of significant interest, is also publicly available.

The NRC modified these rules in 1992 and 2000 to delete reporting requirements for some events that were determined to be of little or no safety significance.   The modified rules continue to provide the Commission with reports of significant events for which the NRC may need to act to maintain or improve reactor safety, or to respond to heightened public concern.   The modified rules also better align requirements on event reporting with the type of information that the NRC needs to carry out its safety mission.   The NRC issued NUREG-1022, Revision 2, "Event Reporting Guidelines, 10 CFR 50.72 and 50.73," in October 2000, concurrent with the rule changes.

NUREG-1022 is structured to help licensees promptly and completely report specified events and conditions.   It discusses general issues that have been difficult to implement in the past, such as engineering judgment, time limits for reporting, multiple failures and related events, deficiencies discovered during licensee engineering reviews, and human performance issues. It also includes a comprehensive discussion of each reporting criterion with illustrative examples and definitions of key terms and phrases.

Event reporting under these rules since 1984 has contributed significantly to focusing the attention of the NRC and the nuclear industry on the lessons learned from operating experience to improve reactor safety.   Over the years, improvements in reactor safety system performance and decreasing trends in the number of reactor transients and significant events have been evident.   Between 2007 and 2010, there were no significant reactor events (defined as having a conditional core damage probability greater than 1×10^{-4}).

Since 2001, the NRC has reviewed each reported reactor-related event and assigned a rating of 1 through 7 on the International Nuclear and Radiological Event Scale.   The agency submits events with a rating of 2 or higher to the IAEA nuclear events Web-based system for public posting.   Other events whose ratings are specifically requested by other member states are also considered for posting regardless of the International Nuclear and Radiological Event Scale rating.   The NRC describes this process in RIS 2002-01, "Changes to NRC Participation in the International Nuclear Event Scale," dated January 2002, and IN 2009-27, "Revised International Nuclear and Radiological Event Scale User's Manual," dated November 2009.

19.7 Programs To Collect and Analyze Operating Experience

As outlined in GL 82-04, "Use of INPO See-in Program," dated March 1982, INPO and the individual licensees are jointly responsible for compiling and analyzing operating experience within the industry.   The effectiveness of licensee operating experience programs is subject to NRC inspection under IP 71152.

The NRC revised its Operating Experience Program in 2005 in response to the recommendations of the Reactor Operating Experience Task Force, established in response to the findings of the Davis Besse Lessons Learned Task Force. Upon launching the revised Operating Experience Program, the NRC implemented some recommendations for better defined roles and responsibilities, a central clearinghouse, and improved collection, storage, and retrieval of information on operating experience. The program process has four phases: (1) collection, (2) screening, (3) evaluation, and (4) application of operating experience data, with a common theme of communication running throughout.

The NRC facilitates the collection, storage, and retrieval of operating experience data with the Operating Experience Gateway, a centralized repository of links to databases relevant to operating experience on the NRC internal Web site, including event reports, international reports, and inspection findings. A database currently under development will provide the same type of centralized data storage and retrieval options for lower level operating experiences, which can be a useful source of information for long-term trending and analysis even though they do not rise to the threshold of reportable events.

The NRC's clearinghouse for operating experience screens event notifications and lower level operating experience from resident inspector feedback to the regional offices daily to determine the level of followup required by each item. The clearinghouse also considers licensee event reports, reports of defects and non-compliance under 10 CFR Part 21, "Reporting of Defects and Noncompliance," international operating experience received from the International Nuclear and Radiological Event Scale Web site and from the IAEA incident reporting system, and any items of potential interest brought forward by the Office of New Reactors and the Office of Nuclear Regulatory Research.

The purpose of 10 CFR Part 21 is to ensure that the NRC receives prompt notification of potential facility, activity, or component deviations or failure to comply, which could cause a substantial safety hazard in facilities or activities licensed by the NRC. These reports are submitted by licensees or vendors to the NRC within sixty days of discovery of the deviation either through the NRC document control desk, or through the NRC Operations Center if it is related to a 10 CFR 50.72 event notification or a 10 CFR 50.73 licensee event report. The reports are forwarded to the operating experience program staff, which conducts the initial assessment and posting of these notifications to the NRC public Web page. If additional information is needed to complete the assessment, the NRC operating experience or vendor quality staff collaborates with the appropriate technical organizations and contacts the submitter of the 10 CFR Part 21 report to obtain more information. The vendor quality staff may also conduct followup actions or inspections with the involved vendor(s). The 10 CFR Part 21 reports are screened by the NRC's clearinghouse for operating experience as described above.

For items that are screened out, followup actions can include e-mail notifications to technical review groups of low-level items for trending and analysis or an operating experience communication distributed internally throughout the agency summarizing the issue and its significance. Items that meet the screening criteria of being both safety significant and generically applicable are screened in as "issues for resolution" (the term used to describe the evaluation phase of the process). Evaluation of an issue for resolution involves an examination of the technical aspects of the issue, and its potential safety significance, as well as an evaluation of previous operating experience.

Finally, the operating experience program applies the results of the evaluation of an issue for resolution.   Application may include the issuance of a generic communication, a proposal for rulemaking, a referral for further study as a generic safety issue, or a revision of IPs.

The NRC participates in the International Nuclear and Radiological Event Scale and the IAEA incident reporting system to both communicate operating experience internationally and review events posted by other member States.   Operating experience personnel review all reactor event notifications received by the agency and rate them on the International Nuclear and Radiological Event Scale.   As Section 19.6 of this report discusses, events with a rating of 2 or higher are posted to the International Nuclear and Radiological Event Scale Web site within 48 hours.   All international events posted to this Web site are screened by the NRC's clearinghouse, as possible issues for resolution based on safety significance and applicability to U.S. plants.   The clearinghouse uses the same criteria to screen the IAEA incident reporting system reports as they are posted.   The NRC submits all U.S. reactor-related generic communications to the IAEA incident reporting system for communication to the international community.

19.8   Radioactive Waste

The NRC has regulations and guidance for nuclear power reactor licensees to ensure the safe management and disposal of low-level radioactive waste.   Onsite low-level waste must be managed in accordance with the NRC regulations in 10 CFR Part 20 and 10 CFR Part 50.   For example, 10 CFR Part 20, Subpart K, "Waste Disposal," deals with licensee treatment and disposition of radioactive waste.   In addition, GL 1981-38, "Storage of Low-Level Radioactive Wastes at Power Reactor Sites," dated November 10, 1981, provides guidance on measures for ensuring the safe storage of low-level waste.

Notwithstanding these regulations and guidance, the economics of waste disposal in the United States have encouraged practices to minimize radioactive waste.   In the past decade or so, disposal costs have risen significantly, and volumes of waste produced have decreased greatly as operations technology evolves.   In June 2008, the NRC published RG 4.21, "Minimization of Contamination and Radioactive Waste Generation: Life-Cycle Planning."   Currently, nuclear power reactors generate only small amounts (about 1,000-2,000 cubic feet per unit) of operational waste each year.

For storage, waste is conditioned into a form that is stable and safe to minimize the likelihood that it will migrate (e.g., as it would if it were a liquid).   Waste that is placed into storage is in a form that is suitable for disposal, or at least a form that can be made suitable for future disposal. The NRC maintains specific regulations for the independent storage of spent nuclear fuel, high-level radioactive waste, and reactor-related low-level waste greater than Class C in 10 CFR Part 72 and detailed regulations for designing and operating low-level waste disposal facilities in 10 CFR Part 61, "Licensing Requirements for Land Disposal of Radioactive Waste."

The U.S. Government addresses in detail the spent fuel and radioactive waste programs, including high-level waste, in a report prepared to satisfy the reporting requirements of the Joint Convention on the Safety of Spent Fuel Management and on the Safety of Radioactive Waste Management.   The latest report (DOE/EM-0654, Revision 2, "United States of America Third National Report for the Joint Convention on the Safety of Spent Fuel Management and on the Safety of Radioactive Waste Management," dated October 2008) is available on the DOE Environmental Management Web site.   In June 2008, DOE submitted a license application to the NRC for the construction of a high-level waste repository at Yucca Mountain, NV.   However, in March 2010, DOE filed a motion to withdraw its application from NRC review.   Concurrently, at the direction of the President of the United States, DOE established the Blue Ribbon Commission on America's Nuclear Future to comprehensively review policies for managing the back end of the nuclear fuel cycle, including all alternatives for the storage, processing, and disposal of civilian and defense used nuclear fuel and nuclear waste.   The Blue Ribbon Commission is expected to make final recommendations to DOE by January 2012.   The NRC will continue to ensure the safe storage of civilian high-level waste.

PART 3

Convention on Nuclear Safety Report:

The Role of the Institute of Nuclear Power Operations in Supporting the United States Commercial Nuclear Power Industry's Focus on Nuclear Safety

August 2010

1. Executive Summary

Following the event at Three Mile Island, the U.S. nuclear power industry established the Institute of Nuclear Power Operations (INPO) in 1979 to promote the highest levels of safety and reliability (i.e., to promote excellence) in the operation of its nuclear power plants. The Institute is a nongovernmental corporation that operates on a not-for-profit basis. Under the United States (U.S.) tax law, the company is classified as a charitable organization that "relieves the burden of government."

Since its inception, all organizations that have direct responsibility and legal authority to operate or construct commercial nuclear plants in the U.S. have maintained continuous membership in the Institute, which currently has 26 members. In addition, many organizations that jointly own these nuclear power plants are associate members. A number of international utility organizations and major supplier organizations also voluntarily participate in the Institute's activities and programs.

In forming INPO, the nuclear utility industry took an unusual step. The industry placed itself in the role of overseeing INPO activities, while at the same time endowing INPO with ample authority to bring pressure for change on individual members and the industry as a whole. This feature makes INPO unique. The industry clearly established and accepted a form of self-regulation through peer review by helping to develop and then committing to meet INPO performance objectives and criteria (POCs). The industry's recognition that all nuclear utilities are affected by the action of any one utility motivated its support of INPO. Each individual member is solely responsible for the safe operation of its nuclear plants. The U.S. Nuclear Regulatory Commission (NRC) has statutory responsibility for overseeing the licensees and verifying that each licensee operates its facility in compliance with Federal regulations to ensure public health and safety. INPO's role -- encouraging the pursuit of excellence in the operation of commercial nuclear power plants -- is complementary but separate and distinct from the role of the NRC.

The nuclear industry's commitment to go beyond regulatory compliance and continually strive for excellence, with INPO's support, has resulted in substantial performance improvements over the last 30 years. For example, in the early 1980s the typical nuclear plant had a capacity factor of 63 percent, experienced six automatic scrams per year, had high collective radiation dose, and experienced numerous industrial safety accidents among its staff. Today, median industry capacity factor is above 91 percent, most plants have no automatic scrams per year, and collective radiation dose and industrial accident rates are both lower by a factor of 7 when compared to the rates of the 1980s.

This report is intended to provide an understanding of the Institute's role and its major programs in support of the U.S. commercial nuclear power industry.

2. Organization and Governance

In many ways, the Institute's organizational structure is similar to a typical U.S. corporation. A Board of Directors, composed of senior executives from INPO's member organizations, provides overall direction for the Institute's operations and activities. Currently, the Board consists of 13 chief executive officers (CEOs) and one president from the member utilities. The Institute's bylaws specify that at least two directors must have recent experience in the direct supervision of operation of a facility that generates electricity or steam for commercial purposes through the application of nuclear power. Also, at least one director must

174

represent a public utility. The president and CEO of the Institute, normally a single individual, is elected by and reports to its Board of Directors. An organization chart is presented below.

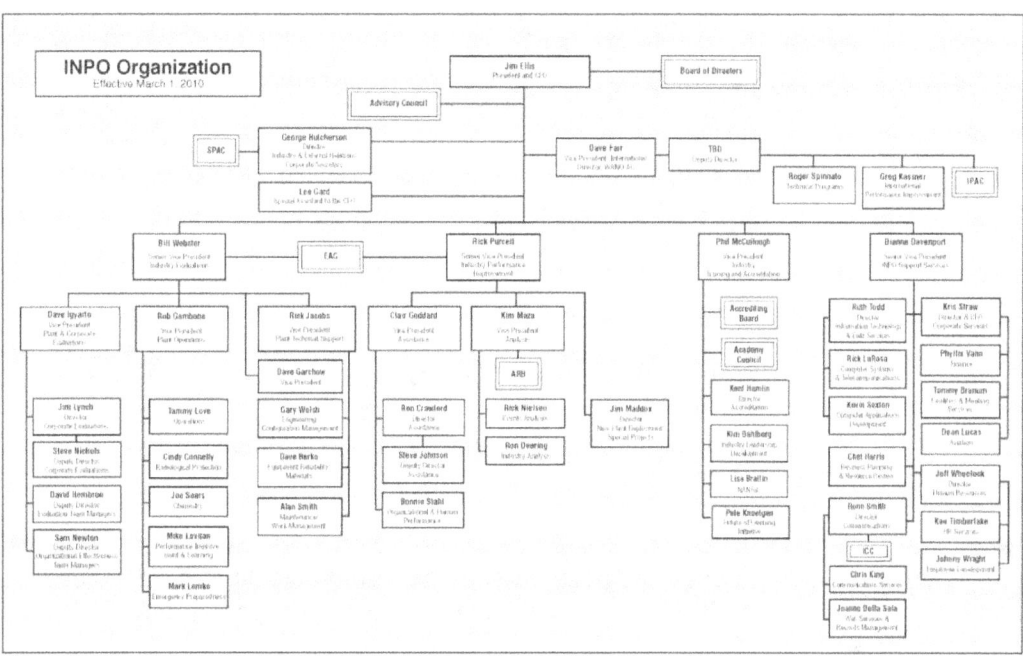

Because the INPO Board of Directors is made up of utility executives, the industry believes that it is important to also have support from an Advisory Council of distinguished individuals, mainly from outside the nuclear generation industry, to provide diversity of experience and thought. This Advisory Council of 9 to 15 professionals selected from outside INPO's membership meets periodically to review Institute activities and provide advice on broad objectives and methods to the Board of Directors. Members include prominent educators, scientists, engineers, and business executives, as well as experts in organizational effectiveness, human relations, and finance.

Institute activities to enhance nuclear plant safety and reliability are reflected primarily in its four cornerstone programs: periodic onsite evaluations of each nuclear plant and corporate support organizations, training and accreditation, events analysis and information exchange, and assistance. Nuclear technical divisions are organized to carry out the cornerstone functions. Other functional areas, such as support services, industry and external relations, and communications, support the nuclear technical divisions as well as the Institute's overall mission.

The National Academy for Nuclear Training operates under the direction of INPO and integrates the training efforts of all U.S. nuclear utilities, the activities of the National Nuclear Accrediting Board, and the training-related activities of the Institute. An INPO executive serves as the executive director of the Academy.

Non-U.S. nuclear organizations from 18 different countries or provinces participate in the Institute's International Participant Program, managed by the World Association of Nuclear Operators (WANO)-Atlanta Centre at INPO's request. This program involves the active exchange of information on nuclear plant operations among utility organizations around the world. Each international participant organization is represented on an advisory committee that provides advice on the operation of this program as well as input on other Institute programs as appropriate. An INPO executive serves as the director of WANO – Atlanta Centre.

Organizations engaged in providing commercial design, engineering, nuclear fuel cycle, or other services directly related to the construction, operation, or support of nuclear electric generating plants also participate in INPO through the Supplier Participant Program. This program allows supplier organizations to share experience and expertise with Institute members and provides a means to provide feedback on operational experience to the suppliers. Currently, 22 companies from around the world are involved in the Supplier Participant Program.

The industry actively participates in the oversight of INPO's programs. Representatives from member utilities serve on the Executive Advisory Group, the Academy Council, the Analysis Review Board, and the Industry Communications Council. The Executive Advisory Group, which consists of the chief nuclear officers of all of the member organizations, advises INPO management on the programs and products in the nuclear technical areas. The Academy Council provides advice in the areas of training, accreditation, and human performance. The Analysis Review Board advises INPO on analysis activities, and the Industry Communications Council advises on effective communication of INPO programs and activities. Frequently, INPO establishes ad hoc industry groups to provide input on specific initiatives.

Financial and Human Resources

The 2010 operating budget for INPO was $95 million, primarily funded through member dues. Dues, approved annually by the Board of Directors, are assessed based on the number of each member's nuclear plant sites and units.

The Institute's permanent staff of about 340 is augmented extensively by industry professionals who serve as loaned employees or international liaison engineers on assignments of typically 18 to 24 months. Loaned and liaison employees comprise about one-third of the total technical staff. They gain extensive experience and training while providing current industry expertise and diversity of thought and practices. A small number of permanent Institute employees serve in loaned assignments to member organizations, primarily for professional development. The total number of both permanent and loaned employees is approximately 400 people.

Institute resources and capabilities are further enhanced by the extensive use of U.S. and international utility peers and executive industry advisors. These peers participate in a wide range of short-term activities, especially on evaluation and accreditation teams that visit nuclear plants. Peers enhance the effectiveness of the INPO teams by offering varied perspectives and providing additional current experience. The peers benefit from learning other ways of conducting business that can be shared with their stations. In 2009, the industry provided INPO with more than 650 peers for short term assignments.

3. INPO's Role within the Federal Regulatory Framework

The nuclear utility industry in the United States, like other industries that may affect the health and safety of the general public, is regulated by the Federal Government. This regulatory function is based principally on the Atomic Energy Act of 1954, as amended, and is carried out by the NRC. In 1979, following the accident at Three Mile Island Nuclear Station, the President of the United States appointed a commission to investigate the accident. The commission, which came to be known as the Kemeny Commission, helped influence the industry's decision to create INPO as a method of self-regulation.

The industry created INPO to provide the means whereby the industry itself could, acting collectively, improve the safety and reliability of nuclear operations. Industry leaders envisioned that peer reviews and POCs based on excellence would be effective in bringing about improvements. In the broad sense, the ultimate goals of the NRC and INPO are the same in that both organizations strive to protect the public; therefore, both review similar areas of nuclear power plant operations. In granting INPO its not-for-profit status, the U.S. Government acknowledged that INPO's role reduces the burden on the Government through the conduct of its activities. However, the industry does not expect INPO to supplant the regulatory role of the NRC. It was recognized that in establishing and meeting its role, INPO would have to work closely with the NRC while at the same time not becoming or appearing to become an extension of or an advisor to the NRC, or an advocacy agent for the utilities. As recognition of their different roles but common goals, the NRC and INPO have entered into a Memorandum of Agreement that includes coordination plans that cover specific areas of mutual interest.

The conduct of plant and corporate evaluations is one of INPO's most important functions. It is also the function that is closest to the role of a regulator. While the two roles -- evaluation and regulation -- may appear similar, they do differ in some ways. The industry and INPO jointly develop numerous POCs. INPO then conducts regular, extensive, and intrusive evaluations to determine how well they are being met. These performance objectives are broad statements of conditions that reflect a higher level of overall plant performance— striving for excellence and often exceeding regulatory requirements. These performance objectives, by their very nature, are difficult to achieve consistently.

Because of the differences in the roles of INPO and the NRC, the industry maintains a clear separation between INPO evaluations and NRC inspections. The industry expects INPO to keep the NRC apprised of its generic activities. While INPO interactions with an individual member remain private between that member and INPO, stations are encouraged to make their INPO plant evaluation and accreditation results available to the NRC for review at each utility or site.

The industry recognizes the need for the NRC to assess the overall quality of INPO's products and the success of its programs. Therefore, the industry expects INPO to provide the NRC with information on INPO programs and activities, including the following:

- copies of selected generic documents
- access to other pertinent information, such as the Equipment Performance Information Exchange (EPIX) database, as described in specific agreements
- observation of certain INPO field activities by NRC employees, with agreement from members

- observation of National Nuclear Accrediting Board sessions

INPO regularly participates in industry-led working groups and task forces that interface with the NRC on specific regulatory issues and initiatives relative to the Institute's mission and strategic objectives. These cooperative interactions have led to the elimination of some redundant activities, benefiting INPO members while enabling both the NRC and INPO to maintain or strengthen focus on their respective missions. For example, the Consolidated Data Entry System, operated by INPO, collects operating data that the NRC uses in its industry oversight process.

INPO has implemented a policy and appropriate procedures with regard to the handling of items that are potentially reportable to the NRC. INPO's policy is to inform utility management of such items during the normal course of business so that the utility can evaluate and report the items as appropriate. If INPO becomes aware of a defect or failure to comply that requires a report under Federal regulation, the Institute has an obligation to ensure that the item is reported, if the utility has not already done so.

4. Responsibilities of INPO and Its Members

INPO members are expected to strive for excellence in the operation of their nuclear plants, to meet INPO performance objectives, and to meet the intent of INPO guidelines. This effort also includes the achievement and maintenance of accredited training programs for personnel who operate, maintain, and support their nuclear plants. Members are expected to be responsive to all areas for improvement identified through INPO evaluation, accreditation, and events analysis programs.

A special procedure, approved by the INPO Board of Directors, provides guidance if a member is not responsive to INPO programs, is unwilling or unable to take action to resolve a significant safety issue, has persistent shortfalls in performance, or has accreditation for its training programs put on probation or withdrawn by the National Nuclear Accrediting Board. The procedure specifies that INPO and the member's management work to resolve any issues in contention using a graduated approach of increasing accountability. Specific options for accountability include interactions between INPO's CEO and the member's CEO and, if necessary, the member's Board of Directors. One option also includes suspending INPO membership if the member continues to be unresponsive. Suspension of membership has never been needed but would have a significant impact on the utility's continued operation, including limiting its ability to obtain insurance.

Furthermore, members are expected to participate fully in other generic INPO programs designed to enhance nuclear plant safety and reliability industrywide. Examples include providing INPO with detailed and timely operating experience information and participating fully in the loaned employee, peer evaluator, and WANO performance indicator programs. Members share information, practices, and experiences to assist each other in maintaining high levels of operational safety and reliability.

In return, INPO is expected to provide members with results from evaluation, accreditation, and review visits, including written reports and an overall numerical assessment that characterizes performance relative to standards of excellence. The industry expects INPO to follow up and verify that effective corrective actions are implemented.

There is clear understanding between INPO and its members that all parties must maintain the confidentiality of INPO evaluation reports and related information, including not distributing this information external to the member utility organization. Members and participants are also expected to use information provided by the Institute to improve nuclear operations and not for other purposes, such as to gain commercial advantage. Members avoid involving INPO or INPO documents in litigation.

INPO members that are also members of the collective insurance organization, Nuclear Electric Insurance Limited (NEIL), have authorized and instructed INPO to make available to NEIL copies of INPO evaluation reports and other data at the Institute's office. NEIL reviews these reports and data for items that could affect the insurability of its members.

INPO POCs are written with input from and the support of the industry. However they are written without regard to constraints or agreements, such as labor agreements, of any individual member. Each member is expected to resolve any impediments to their implementation that may be imposed by outside organizations.

INPO does not engage in public, media, or legislative activities to promote nuclear power. Such activities would undermine INPO's objectivity and credibility and may jeopardize the Institute's not-for-profit status.

5. Principles of Sharing (Openness and Transparency)

Throughout the changes that have occurred in the U.S. electric industry, including the process of electric deregulation, the industry has reaffirmed INPO's mission to promote the highest levels of safety and reliability (i.e., to promote excellence) in the operation of nuclear power plants. Even with U.S. utilities now in competition in certain areas, there is a clear understanding of the need to continue sharing pertinent operational information to continuously strengthen safety and reliability. Nuclear utility owners believe that this cooperation is fundamental to the industry's continued success.

Through INPO, nuclear utilities quickly share information important to safety and reliability, including operating experience, operational performance data, and information related to failure of equipment that impacts safety and reliability. The industry also actively encourages benchmarking visits to support the sharing of best practices and the concepts of emulation and continuous improvement.

INPO facilitates industry information sharing by including participation of industry peers in the INPO cornerstone programs—plant evaluations, training and accreditation, analysis and information exchange, and assistance. INPO communicates and shares information through a variety of methods, including the secure member Web site, Nuclear Network®, written guidelines, and other publications.

While the industry and INPO recognize that rapid and complete sharing of information important to nuclear safety is essential, there is a clear understanding that certain information is private in nature and is not appropriate to share. Examples are INPO plant-specific details of evaluation and accreditation results, personal employee and individual performance information, and appropriate cost and power marketing data.

6. Priority to Safety (Safety Culture)

The U.S. nuclear industry believes that a strong safety culture is central to excellence in nuclear plant operations, partly because of the special and unique nature of nuclear technology and the associated hazards—radioactive byproducts, concentration of energy in the reactor core, and decay heat. Within our members' power plants and within INPO, the elements, activities, and behaviors that are part of a strong safety culture are embedded in everything that we do day to day and have been since INPO was formed in 1979.

> *The U.S. nuclear industry has defined safety culture as follows: An organization's values and behaviors—modeled by its leaders and internalized by its members—that serve to make nuclear safety the overriding priority.*

To support line managers in fostering a strong safety culture, the nuclear industry developed the *Principles for a Strong Nuclear Safety Culture* in November 2004. The principles were incorporated into the POCs as the foundation of nuclear safety in May 2005. The following eight principles are the foundation of a strong nuclear safety culture:

1. Everyone is personally responsible for nuclear safety.

2. Leaders demonstrate commitment to safety.

3. Trust permeates the organization.

4. Decision-making reflects safety first.

5. Nuclear technology is recognized as special and unique.

6. A questioning attitude is cultivated.

7. Organizational learning is embraced.

8. Nuclear safety undergoes constant examination.

INPO activities reinforce the primary obligation of the operating organizations' leadership to establish and foster a healthy safety culture, to periodically assess safety culture, to address shortfalls in an open and candid fashion, and to ensure that everyone from the board room to the shop floor understands his or her role in safety culture.

As part of its focus on safety, the industry utilizes INPO, through evaluations and other INPO activities, to identify and help correct early signs of decline in safety culture at any plant or utility. Further, the industry has defined INPO's role as follows:

- Define and publish standards relative to safety culture.
- Evaluate safety culture at each plant.
- Develop tools to promote and evaluate safety culture.
- Assist the industry in providing safety culture training.
- Develop and issue safety culture lessons learned and operating experience.
- Make safety culture visible in various forums such as professional development seminars, assistance visits, working meetings, and conferences including the CEO conference.

In 2002, INPO published Significant Operating Experience Report (SOER) 02-4, "Reactor Pressure Vessel Head Degradation at Davis-Besse Nuclear Power Station." The purpose of the report was to describe the event and the shortfalls in safety culture that contributed to the event, as well as to recommend actions to prevent similar safety culture problems at other plants. This event is considered a defining moment in the U.S. nuclear power industry, highlighting problems that can develop when the safety culture at a plant receives insufficient attention. Every U.S. nuclear power station has implemented the SOER recommendations, and INPO evaluation teams have reviewed each station's actions. Briefly, the recommendations encompass discussing a case study on the event with all managers and supervisors in the nuclear organization, periodically conducting a self-assessment to determine the organizational respect for nuclear safety, and identifying and resolving abnormal plant conditions or indications that cannot be readily explained. This SOER has also been shared with WANO and republished as a WANO document.

Safety culture is thoroughly examined during each plant evaluation. Each evaluation team is expected to evaluate safety culture throughout the process, including during the pre-evaluation analysis of plant data and observations made at the plant. The results of this review are included in the summary on organizational effectiveness and may be documented as an area for improvement, as appropriate. The INPO evaluation team discusses aspects of a plant's safety culture with the CEO of the utility at each evaluation exit briefing.

In February 2009, INPO proposed aligning the language used by INPO and the NRC when describing safety culture. In June 2009, leadership from the NRC and INPO met to discuss the possibility of this happening and define high-level expectations. In December 2009, the NRC announced a series of meetings, planned for 2010, where a selected panel of stakeholders would jointly craft a high-level definition of safety culture and identify/define the major components within safety culture.

Also in 2009, and in response to industry requests, INPO developed an addendum to the *Principles for a Strong Nuclear Safety Culture*. This addendum lists specific behaviors that are indicative of a strong nuclear safety culture. These behaviors are more specific than those listed in the *Principles for a Strong Nuclear Safety Culture* and are arranged by organizational level, from senior managers to individual contributors.

7. **Cornerstone Activities**

   a. **Evaluation Programs**

      Members host regular INPO evaluations of their nuclear plants approximately every 2 years. Additional evaluative review visits are periodically conducted on corporate support and other more specific areas of plant operation. During these evaluations and reviews, the INPO teams use standards of excellence based on the POCs and their own experience, as well as their broad knowledge of industry best practices. This approach shares beneficial industry experience while promoting excellence in the operation, maintenance, and support of operating nuclear plants. Written POCs, developed by INPO with industry input and review, guide the evaluation process and are the bases for identified areas for improvement. The evaluations are performance oriented, emphasizing both the results achieved and the behaviors and organizational factors important to future performance. The evaluations focus on those issues that impact nuclear safety and plant reliability.

181

i. Plant Evaluations

Teams of approximately 15 to 20 qualified and experienced individuals conduct evaluations of operating nuclear plants, focusing on plant safety and reliability. In 2009, U.S. utilities received 38 plant evaluations or WANO peer reviews. The evaluation teams are augmented by senior reactor operators, other peer evaluators from different utilities, host utility peer evaluators, and an executive industry advisor. The scope of the evaluation includes the following functional areas:

- operations
- maintenance
- engineering
- radiological protection
- chemistry
- training

In addition, teams evaluate cross-functional performance areas (i.e., processes and behaviors that cross organizational boundaries) and address process integration and interfaces. The following cross-functional areas are evaluated:

- safety culture
- operational focus
- configuration management
- equipment reliability and work management
- performance improvement (learning organization)
- organizational effectiveness

Team leaders, in addition to leading and coordinating team activities, provide a focal point for evaluation of station management and leadership, concentrating on evaluating leadership, organizational effectiveness, safety culture, and nuclear oversight topics.

The performance of operations and training personnel during simulator exercises is included as a key part of each evaluation. Also included, where practicable, are observations of refueling outages, plant startups, shutdowns, and major planned evolutions.

The evaluation team provides the utility with formal reports of strengths and areas for improvement, along with a numerical rating of overall plant performance. As part of the 1983 annual INPO CEO workshop, INPO prepared a set of indicators for each nuclear station that reflected station participation in and commitment to INPO programs. INPO provided this information to each CEO. One of these indicators was an assessment of each station's overall performance based on INPO evaluations and the judgment of INPO team managers and senior management.

With the approval of the Board of Directors, INPO decided that an assessment of overall station performance in the context described above would be made after each evaluation and shared privately with the CEO at the exit meeting. Eventually a numerical assessment was developed, and each station is now provided an assessment from category 1 (Excellent) to 5, which is defined as a level of performance where the margin to nuclear safety is substantially reduced. Such a process reflects the desire of utility managers to know more precisely how their station's performance compares relative to the standards of excellence. It is also in keeping with INPO's responsibility to the individual CEO and to its members for identifying low-performing nuclear plants and for stimulating improvement in performance.

Even though standards for performance have risen substantially over the years, the number of plants in categories 1 and 2 has remained relatively constant, even as standards of excellence have improved. Additionally, several conclusions can be drawn from evaluations over the years. Excellent plants (category 1) and category 2 plants show strong leadership, are self-critical, do not tolerate complacency, are operationally focused, have exceptional equipment performance, and effectively use training to improve performance. Attributes of category 3 and 4 stations may include leaders not setting high standards, a weak self-critical attitude, weak day-to-day operations, broad equipment problems, and deficient fundamental knowledge and skills in several areas. It has been over a decade since a station has been assessed in category 5.

The final report includes utility responses to the identified areas for improvement, along with their commitments to specific corrective action. In subsequent evaluations and other interactions, INPO specifically reviews the effectiveness of actions taken to implement these improvements.

In addition to the strengths and areas for improvement provided in the evaluation report, subjective team comments are often communicated to the member CEO during the evaluation exit meeting. These comments, often more intuitive, are intended to help utilities recognize and address potential issues before they adversely affect actual performance. Copies of the plant evaluation report are distributed according to a policy approved by the Institute's Board of Directors.

The industry also hosts WANO peer reviews conducted by the WANO-Atlanta Centre. These are conducted at each U.S. station approximately every 6 years and are performed in lieu of an INPO plant evaluation at each station. These peer reviews use a methodology similar to that of plant evaluations, but with teams augmented with international peers.

Numerous improvements have been made in plant safety and reliability as a result of addressing issues identified during evaluations, peer reviews, plant self-assessments and comparison and emulation among plants. The time plants operate versus the amount of time they are shutdown has improved significantly, the frequency of unplanned shutdowns has decreased markedly, and the reliability and availability of safety systems has improved measurably.

ii. **Corporate Evaluations**

Member utilities that operate multiple nuclear stations request that INPO conduct corporate evaluations on an interval of 4 to 6 years. Corporate evaluations at single nuclear station utilities are conducted when requested by the utility or when deemed necessary by INPO. The INPO-conducted corporate evaluations reflect the important role of the company headquarters in supporting the successful operation of plants within a multi-site fleet. INPO conducted five corporate evaluations in 2009.

A tailored set of POCs define the scope of activities and the standards for corporate evaluations. The corporate evaluation focuses on the impact that the corporation has on the safe operation of its nuclear plants. Areas typically evaluated during a corporate evaluation include the following:

- direction and standards for station operation, including the organizational alignment, communications, and accountability for strategic direction, business and operational plans, and performance standards
- governance, monitoring and independent oversight of the nuclear enterprise
- support for emergent station issues and specialty areas such as major plant modifications, including replacement of steam generator and reactor vessel heads and station upgrades to extract more power and efficiency
- performance of corporate functions, such as human resources, industrial relations, fuel management, supply chain management and other areas, as applicable to the nuclear organization

INPO members use corporate evaluation results to help ensure that essential corporate functions are providing the leadership and support necessary to achieve and sustain excellent nuclear station performance. As a consequence of responding to issues identified during corporate evaluations, appropriate resources and leadership attention have often been refocused on improving station safety and reliability.

At the request of its members, INPO meets with utility boards of directors to provide an overview of plant, and when applicable, fleet performance. These briefings are used by the boards of directors as an input to their assessment of operational risk.

iii. Other Review Visits

The industry also utilizes INPO to conduct review visits in selected industrywide problem areas to supplement the evaluation process. These visits are typically initiated by INPO and are evaluative in nature. The results of review visits may be used as an input to the evaluation process. The visits are designed as in-depth reviews of technical areas that could have a significant impact on nuclear safety and reliability. Such areas include critical materials issues that affect the structural integrity of the reactor coolant system and reactor vessel internals of both boiling-water reactors (BWRs) and pressurized-water reactors (PWRs). Other areas include components or systems that are significant contributors to unplanned plant transients and forced loss rate, including main generator and transformer, switchyard, and electrical grid components. In 2009, INPO conducted 109 review visits.

Similar to plant evaluations and peer reviews, review visits evaluate station performance against the INPO POCs to a standard of excellence. In some areas, such as materials, industry groups have developed detailed technical guidance that each utility has committed to implement. The materials review visit teams also use

184

this guidance to ensure that program implementation is consistent and complete and meets the industry-developed standards.

Review visit teams are led by an INPO employee and include industry personnel who have unique expertise in the area of the review that is not typically within the skill set of INPO members of plant evaluation or peer review teams. Review visits typically include a week of preparation followed by a week on site.

Review visit reports contain beneficial practices and recommendations for improvement. These reports are sent to the station site vice president. For potential safety-significant recommendations, INPO may request a response. The subsequent plant evaluation or WANO peer review team follows up on each of the recommendations that require a response to ensure that identified issues are addressed. Periodically, INPO compiles the beneficial practices and recommendations and posts the information on the secure member Web site to allow all utilities to benchmark their programs.

The following sections discuss the details of selected review visit programs.

Pressurized-Water Reactor Steam Generator Review Visits

INPO initiated steam generator review visits in 1996. In the early 1980s, steam generator tube leaks and ruptures were significant contributors to lost power generation and were the cause of several events deemed significant by INPO. The industry as a whole became more sensitive to the importance of steam generator integrity as a contributor to core damage frequency analysis. The industry, through the Electric Power Research Institute (EPRI) Steam Generator Management Program, developed and maintained detailed guidance on qualification and implementation of nondestructive testing techniques, engineering assessments of steam generator integrity, and detection and response to tube leakage and ruptures. In mid-1995, the industry requested that INPO help improve the prevention and detection of steam generator degradation by verifying correct and consistent implementation of industry guidance at individual stations and to evaluate steam generator management programs against standards of excellence. As a result, INPO established the steam generator review visit program. Other review visits that were initiated later used the steam generator review visit process as a model.

Steam generator review visits focus on steam generator in-service inspection and repair, use of qualified personnel and techniques for eddy-current examinations of tubes; tube plugging procedures; assessment of current inspection results; chemistry conditions that affect steam generators; and steam generator primary-to-secondary leak detection, monitoring, and response.

In general, steam generator management programs have steadily improved and are implemented effectively, as evidenced by the lack of safety-significant events and events that contribute to lost generation. Steam generator replacements have also contributed to overall improved performance. Consequently, steam generator review visits currently identify few significant issues. However, the review visits have identified a need for improved timeliness in implementing industry-developed or revised guidance, and improved rigor in inspecting for, evaluating, and retrieving loose parts.

185

Boiling-Water Reactor Vessel and Internals Review Visits

In 2001, INPO initiated BWR vessel and internals review visits at the request of the industry. In the early 1990s, vessel and internal issues caused by intergranular stress-corrosion cracking became significant contributors to lost power generation. Safety concerns associated with this degradation prompted the industry to form the EPRI BWR Vessel and Internals Project. This group developed detailed guidance to address inspection, mitigation, repair, and evaluation of degradation for components important to safety and reliability.

BWR vessel and internals review visits focus on nondestructive examinations; inspection scope and coverage; evaluation of crack growth and critical flaw size; effectiveness of strategies to mitigate intergranular stress-corrosion cracking, including hydrogen addition and application of noble metals; and chemistry conditions that affect long-term health, including potential effects on fuel.

Industry overall performance has improved as evidenced by the lack of safety-significant events and events that contribute to lost generation.

Pressurized-Water Reactor Primary Systems Integrity Review Visits

INPO initiated PWR primary systems integrity review visits in 2003. Since the early 1980s, a number of notable events associated with leakage from PWR borated systems have resulted in additional oversight by the NRC and INPO. In some cases, these leakage events have resulted in corrosion and wastage of reactor coolant system pressure-retaining components. The EPRI PWR Materials Reliability Program was formed as an industry initiative in 1998 to develop guidance to address materials degradation issues. Because of the importance of primary systems integrity, INPO began performing in-depth review visits focused on boric acid corrosion control and Alloy 600 degradation management, including dissimilar metal butt welds.

PWR primary systems integrity review visits focus on the inspection and evaluation of reactor coolant system pressure-retaining components; the qualification of nondestructive examination personnel and techniques; and the monitoring and response to unidentified leakage in containment, including management guidance and operator procedures.

As a result of these industry efforts, performance appears to be improving. Stations are identifying degradation before leakage occurs. Stations have also more aggressively pursued indications of minor unidentified leakage. Alloy 600 dissimilar metal butt weld examinations and mitigation will continue over the next few years as the enhanced industry-defined actions continue to be performed and inspections take full advantage of improved nondestructive examination techniques.

Transformer, Switchyard, and Grid Review Visits

INPO initiated transformer, switchyard, and grid review visits in 2004. Many transformers have been in service for numerous years and are often the original station transformers. Considering this aging—along with the recent trends of power uprates, license renewal, and increased loading—these transformers may be operating with a reduction in margin. With this decrease in margin, the need for increased monitoring, trending, and predictive and preventive maintenance became apparent in order to identify and mitigate potential problems before they result in on-line failure. Additionally, a series of events in 2003, including the blackout in the northeastern United States and parts of Canada, reinforced the need for nuclear plants to have reliable offsite power. There was also renewed focus on how nuclear plant conditions and electrical power system line-ups to the switchyards can help minimize and prevent grid events.

The transformer, switchyard, and grid review visits focus on communication and coordination with grid operators, including formal agreements and implementing procedures, adequacy of offsite power, and predictive and preventive maintenance for large power transformers and switchyard equipment.

While isolated events related to switchyards, transformers, and grids continue to occur, additional rigor in maintenance and interfaces has shown some improvement. Additionally, sharing of information and lessons learned among utilities is resulting in implementation of barriers to prevent future events. It is expected that as the review visits continue, the number and significance of events will be reduced.

Main Generator Review Visits

The industry initiated main generator review visits were in 2004 following identification of an adverse trend involving failures of main generators and related support systems. The number of main generator failures that hindered power production or extended an outage, or both, had doubled from 1999 to 2003. During this time, unplanned scrams caused by generator problems increased to around five per year from the previous average of two per year. The most frequent generator maintenance challenges involved support systems, such as stator cooling water and the exciter, and often included human performance elements. As a result of industry identification of this adverse performance, INPO began conducting main generator review visits to focus on improving the performance of main generators.

Main generator review visits focus on performance and condition monitoring to ensure that the generator is operating within design parameters and to detect early signs of equipment degradation, preventive and condition-based maintenance to address the effects of aging, outage planning to ensure that important main generator work is performed, and knowledge and skill levels of personnel to ensure proper workmanship.

Emergency Preparedness

In 2007, INPO reestablished its emergency preparedness section to help the industry continue to improve its readiness to respond to radiological and other site emergencies. INPO began this initiative in response to a need identified in 2002 by the Nuclear Energy Institute and a subsequent industry review led by INPO of 25 plants over 3 years. These visits identified opportunities for improvement that included more timely and accurate classifications, notifications, and protective action recommendations; strengthened drill programs; and increases in emergency response organization staffing. The emergency preparedness review visit program is a formal INPO program with each site receiving a visit every 4 years.

In 2010, INPO entered its fourth year of conducting emergency preparedness review visits. During this time, INPO identified several industrywide issues, which are being addressed by working groups comprising industry leaders and facilitated by INPO. INPO developed and published a guideline that provides a basic task analysis and training program elements for key emergency response organization members. The Institute is drafting additional guidance on how to better control equipment important to emergency preparedness and on how to develop realistic training and evaluation of shift manager oversight during emergencies. INPO anticipates that published guidance on these topics will be available to the industry in 2010.

INPO also conducted the fourth annual emergency preparedness manager seminar in 2010. As turnover and attrition continues to challenge the industry, demand for qualified emergency preparedness managers spotlights the need for this highly sought after seminar. The 1-week seminar is intended to address this ongoing turnover. Another initiative expected to prove valuable the establishment of periodic industrywide working meetings at INPO. These meetings will address a broad range of industry issues identified by our members and are expected to capitalize on gathering a broad range of experienced program owners to address specific topics.

The INPO Emergency Plan and the recently updated Emergency Response Center is used to assist members in mobilizing the resources of the nuclear industry and to provide other resources or assistance as necessary, following classification of an emergency event. INPO recently completed an emergency response drill, performed with support of an industry fleet emergency preparedness organization. This drill demonstrated the value of a collaborative relationship with industry members in providing needed support.

b. Training and Accreditation Programs

The U.S. commercial nuclear power industry strongly believes that proper training of plant operators, maintenance workers, and other support group workers is of paramount importance to the safe operation of nuclear plants. As a result, the industry established the National Academy for Nuclear Training in 1985 to operate under the responsibility of INPO. The industry formed the Academy to focus and unify high standards in training and qualification and to promote professionalism of nuclear plant personnel. The Academy integrates the training-related activities of all members, the independent National Nuclear Accrediting Board, and the Institute. Through INPO, the Academy conducts seminars and courses and provides other training and training materials for utility personnel.

All U.S. nuclear plants have accredited training programs and are branches of the Academy. A utility becomes a member of the Academy when all of its operating plants have achieved accreditation for all applicable training programs.

INPO interacts with all members in preparing for, achieving, and maintaining accreditation of training programs for personnel involved in the operation, maintenance, and technical support of nuclear plants. These interactions, similar in content to the accreditation efforts of schools and universities, include evaluations of accredited training programs, activities to verify that the standards for accreditation are maintained, and assistance at the request of member utilities. Written objectives and criteria are jointly developed with the industry and guide the accreditation process.

Unlike our role in the plant evaluation and assessment process described above, INPO is not the accrediting agency. The independent National Nuclear Accrediting Board examines the quality of utility training programs and makes all decisions with respect to accreditation. If training programs meet accreditation standards, the Board awards or renews accreditation. If significant problems are identified, the Board may defer initial accreditation, place accredited programs on probation, or withdraw accreditation. Accreditation is maintained on an ongoing basis and is formally renewed for each of the training programs every 4 years. The National Nuclear Accrediting Board, comprised of training, education, and industry experts, is convened and supported by INPO, but it is independent in its decisionmaking authority. Board members are selected from a pool of individuals from utilities, post-secondary education, nonnuclear industrial training, and NRC nominations. Each Board consists of five sitting members, with a maximum of two utility representatives to ensure Board independence from the nuclear industry.

The accreditation process is designed to identify strengths and weaknesses in training programs and to assist in making needed improvements. The process includes self-evaluations by members, with assistance provided by INPO staff; on-site evaluations by teams of INPO and industry personnel; and decisions by the independent National Nuclear Accrediting Board. Members are expected to seek and maintain accreditation of training programs for the following positions or skill areas:

- shift managers
- senior reactor operators
- reactor operators
- nonlicensed operators
- continuing training for licensed personnel
- shift technical advisors
- instrument and control technicians and supervisors
- electrical maintenance personnel and supervisors
- mechanical maintenance personnel and supervisors
- chemistry technicians
- radiological protection technicians
- engineering support personnel

In 2002, the industry updated the accreditation objectives to place additional emphasis on training for performance improvement. It was recognized that in striving for excellence, training must be an integral part of each plant's business strategy and daily operations to ensure a highly skilled workforce. This approach strengthens the link between the analysis of performance gaps and the training that results in tangible improvements in people and plant activities. The five-step systematic approach to training remains the essential tool for providing training that is results oriented. Both line and training organizations are expected to work together to analyze performance gaps and to design, develop, and deliver training that enhances knowledge and skills to measurably improve plant performance. Such an approach to improving worker knowledge and skills contributes to high levels of safety, as seen in industry gains in equipment reliability, safety system availability, collective radiation exposure, and worker

189

safety, as well as fewer events. The role of training will continue to be vital in coming years as many experienced workers retire and new workers enter the workforce.

In 2009, the National Nuclear Accrediting Board renewed accreditation for 164 of 182 training programs. Eighteen programs at three stations were placed on 6-month probation and required to upgrade their training programs. After considerable corrective actions and investment, both stations were successful in having their programs' accreditation renewed following the probation period and after presenting their improvements to the Accrediting Board. The third station will return to the Accrediting Board in 2010.

While the accreditation process is independent of the NRC, it is recognized and endorsed by the NRC as a means for satisfying regulatory training requirements. In its "Annual Report on the Effectiveness of Training in the Nuclear Industry," the NRC noted that, "Monitoring the INPO managed accreditation process continued to provide confidence that accreditation is an acceptable means of ensuring the training requirements contained in 10 CFR 50 and 10 CFR 55 are being met." In addition, the NRC assessment of the accreditation process indicates that continued accreditation remains a reliable indicator of successful systematic approach to training implementation and contributes to the assurance of public health and safety by ensuring that nuclear power plant workers are being trained appropriately.

i. **Training and Qualification Guidelines**

The Academy develops and distributes training and qualification guidelines for operations, maintenance, and technical personnel. These guidelines are designed to assist the utility in developing quality training programs and in selecting key personnel.

Training and qualification guidelines are revised and updated periodically to incorporate changes to address industry needs and to take into account lessons learned from other INPO programs such as evaluations, events analyses, working meetings, and workshops. These training and qualification guidelines provide a sound basis for utility training programs.

ii. **Courses and Seminars**

The industry benefits extensively from courses and seminars that the Academy conducts to help personnel better manage nuclear technology, more effectively address leadership challenges, and improve their personal performance. In 2009, nearly 1,400 industry employees, including many international representatives, participated in more than 70 courses and seminars. Examples of courses and seminars conducted are as follows:

- Goizueta Director's Institute (focused on the directors of member boards) (INPO, in partnership with the Goizueta Business School of Emory University, conducts "The Impact of Governance on the Nuclear Power Industry," a nuclear education course designed for directors in the nuclear industry. Since its inception in 2006, the program has attracted 146 participants from member and international utilities.)
- Reactor Technology Course for Utility Executives

190

- Senior Nuclear Executive Seminar
- Senior Nuclear Plant Management Course
- Human Performance Fundamentals Course
- High Performance Teamwork Development
- Operations Supervisor Professional Development Seminar
- First-Line Leadership Seminar
- Next-Level Leadership Seminar
- Seminars for new plant managers and for new managers in operations, radiological protection, chemistry, maintenance, engineering, nuclear oversight, and training

In February 2006, INPO launched the National Academy for Nuclear Training e-Learning (NANTeL) system. Using Web-based technologies allowing distance learning, NANTeL training includes courses and proctored examinations for plant access, radiation worker, human performance, and industrial safety qualification to industry standards. By July 2006, all member utilities had agreed to participate in the system by accepting generic training and updating the industry's Personnel Access Data System for training course completions. The system offers 42 generic and 215 utility or site-specific training courses. Between March 1, 2006, and December 31, 2009 more than 100,000 industry workers have completed a total of 1,059,840 courses.

Meeting the challenges of developing a well-trained, knowledgeable workforce in the future continues to receive attention. Early in 2008, INPO began work on the first phase of a new industry initiative called the Future of Learning. Developed with extensive industry participation, this initiative lays out a strategy to guide training efforts in the years ahead. It will help the industry deal with workforce renewal, the training of a new generation of workers, and the training of even more workers to support new plant construction.

INPO efforts to help prepare and energize the nuclear workforce of tomorrow include a new leadership seminar designed for emerging nuclear leaders. Also, the "Nuclear Citizenship for New Workers" course, emphasizing the uniqueness of our nuclear industry, has been made available, as well as an industrywide instructor training and certification program that uses a blend of distance learning and classroom instruction.

c. **Analysis and Information Exchange Programs**

The analysis and information exchange programs improve plant safety by identifying the causes of industry events that may be precursors to more serious events. Stations are required to share operating experiences and lessons learned with INPO. INPO then analyzes and rapidly communicates the information to the industry through a variety of methods and products. In addition, INPO analyzes a variety of operational data to detect trends in industry performance and communicates the results to the industry.

INPO operates and maintains extensive computer databases to provide members and participants ready access to information on plant and equipment performance and operating experience. These databases are accessible from INPO's secure member Web site. For example, the industry uses Nuclear Network®, a worldwide internet-based communication system, to exchange information on the safe operation of nuclear plants. WANO also uses Nuclear Network® as a primary means for communicating and exchanging operating experience among its members and regional centers.

i. **Events Analysis Program**

INPO reviews and analyzes operating events from both domestic and international nuclear plants through its Significant Event Evaluation and Information Network (SEE-IN) Program. The program is designed to provide in-depth analysis of nuclear operating experience and to apply the lessons learned across the industry. Events are screened, coded, and analyzed for significance; those with generic applicability are disseminated to the industry in one or more of the following forms, beginning with events of greatest importance:

- SOERs
- Significant Event Reports (SERs)
- Significant Event Notifications (SENs)

Members support the events analysis program by providing INPO with detailed and timely operating experience information. Operating experience information is freely shared among INPO members. The U.S. industry submits more than 2,000 operating experience entries every year, or about 30 to 40 per station. These entries enable a single station to multiply its experience base for identifying problems. This experience base includes safety systems, which have similar components across many stations. For example, one station recently discovered scoring of a cylinder on an emergency diesel generator (EDG) that could render the EDG inoperable. Other stations were able to use this information to take actions to inspect their EDGs before actual equipment malfunction. A key to this success is the timeliness of reporting. Stations typically report events in less than 50 days after occurrence.

Members are required to evaluate and take appropriate action on recommendations provided in SOERs. During on-site plant evaluations, INPO teams follow up on the effectiveness of each station's actions in response to SOER recommendations. For example, during a recent plant evaluation, team members reviewing SOER recommendations identified a potentially significant transformer problem that likely would lead to catastrophic failure if not corrected in a timely manner. This event was avoided because of lessons documented in an SOER. Topics of SOERs in recent years include loss of grid, reactivity management, reactor core designs, transformers, unplanned radiation exposures, and rigging and lifting of heavy loads.

Members should review and take actions as appropriate on SENs, SERs, and other reports provided by INPO. INPO evaluates the effectiveness of utility programs in extracting and applying lessons learned from industrywide, as well as internal station, operating experience.

INPO maintains all operating experience reports since the start of the SEE-IN Program in searchable databases available on the secure member Web site. This information supports members in applying historical lessons learned as new issues are analyzed or activities are planned. INPO also provides "just-in-time" briefing summaries in numerous topical areas in a format designed to help plant personnel prepare to perform specific tasks. These documents provide ready-to-use materials to brief workers on problems experienced and lessons learned during recurring activities.

ii. Other Analysis Activities

INPO analyzes industry operational data from a variety of sources—events, equipment failures, performance indicators, and regulatory reports—to detect trends in industry performance. INPO communicates the results of analyses to the industry using several methods, including topical reports. These documents typically review events and other data over a period of years to summarize performance trends and causes and suggest actions. Subjects of recent topical reports include fuel reliability, foreign material intrusion, intake cooling blockage, large motor failures, and contractor personnel performance. Stations use these reports to assess their performance and identify improvements. In addition, individual plant performance data are analyzed, with results used to support other INPO activities, such as evaluations and assistance.

iii. Nuclear Network® System

Nuclear Network is an international electronic information exchange for sharing nuclear plant information. It is the major communication link for the SEE-IN and WANO event reporting system. The system transmits operating experience information, SERs, and other nuclear technical information.

The system includes a special dedicated method for reporting unusual plant situations. This feature allows the affected utility to provide timely information simultaneously to all Nuclear Network® users, including the U.S. industry, INPO's international and supplier participants, and WANO members, so the affected station does not have to respond to multiple inquiries. In addition, members are promptly informed of problems occurring at one station, allowing them to implement actions to prevent a similar occurrence.

iv. Performance Data Collection and Trending

INPO operates and maintains a consolidated data entry system as a single process by which to collect data and information related to nuclear plant performance. Members provide routine operational data in accordance with the WANO Performance Indicator Program or regulatory requirements on a quarterly basis. These plant data are then consolidated for trending and analysis purposes. Industrywide data, plus trends developed from the data, are provided to member and participant utilities for a number of key operating plant performance indicators. Members use these data for comparison and emulation, in setting specific performance goals, and in monitoring and assessing performance of their nuclear plants.

In the mid-1980s, the industry worked with INPO to establish a set of overall performance indicators focused on plant safety and reliability. These indicators have gained strong acceptance and use by utilities to compare performance, set targets, and drive improvements. Examples of indicators collected and trended include unplanned automatic scrams, safety systems performance, unit capability factor, forced losses of generation, fuel reliability, collective radiation exposure, and industrial safety accidents.

The industry has established long-term goals for each indicator on a 5-year interval, beginning in 1990. Annex 2 of this report provides key performance indicator graphs for U.S. plants.

v. Equipment Performance Data

INPO operates and maintains the EPIX system, which tracks the performance of equipment important to safety and reliability. The industry reports equipment performance information to EPIX in accordance with established guidance. Member utilities use the data to identify and solve plant equipment performance problems, with the goal of enhancing plant safety and reliability. The information is also used by the Institute for performance trending to identify industrywide performance problems. INPO also makes the data available to the NRC to support equipment performance reviews by the regulator.

vi. Operating Experience for New Plant Construction

In 2009, a means for collecting and distributing experience from construction problems was established through the U.S. industry's Nuclear® Network System. Nuclear Network® has long been the forum for rapid and secure communications and has hosted the industry's operating experience program. The new construction experience program has a similar mission to that of the operating experience, but it is tailored to the unique needs of utilities with construction projects.

d. Assistance Programs

Between evaluations, a station can request and receive assistance in specific problem areas to help improve plant performance. In addition, INPO monitors the performance of member utility stations between evaluations to identify areas in which assistance can be used to improve plant performance or respond to declining performance. The purpose of this monitoring is to identify, as early as possible, stations that exhibit indications of declining performance so that focused assistance can be provided to help reverse the performance trend. INPO also provides members with comparisons of their plants' performance to overall industry performance in a variety of areas.

A majority of assistance visits to member utilities by INPO personnel and industry peers are at the request of the stations. This assistance is targeted for specific technical concerns, as well as for broader management and organizational issues. While assistance is generally requested by a station, in some cases INPO may suggest assistance in a specific area to stimulate improvements.

Assistance resources are provided using a graded approach that provides a higher priority to those plants that need greater performance improvement. An INPO management senior representative is assigned to each station to facilitate assistance efforts. Station and utility management maintains close liaison with the senior representative to help identify where INPO resources can best be used to address specific issues and help improve overall station performance.

When significant performance shortfalls persist at a station or when performance trends indicate chronic conditions that could detract from safe and reliable plant operation INPO will follow a policy of graduated engagement with the member utility. For a nuclear plant that shows either consistently poor performance over several evaluation cycles or a significant decline in performance between evaluation cycles, the INPO staff will recommend and obtain concurrence from the INPO CEO to include the plant in a special focus category. For plants that need special focus, INPO will establish a Special Focus Oversight Board that will conduct scheduled periodic reviews to determine the effectiveness of station improvement activities and provide rapid feedback. Board members will usually include both industry and INPO executives.

INPO provides documents that describe nuclear safety principles, effective leadership and management practices, and good work processes and practices to assist member utilities. Members help INPO develop these documents and then use them to address specific improvement needs.

Workshops, seminars, working meetings, and other activities are also conducted to assist in the exchange of information among members and to support the development of industry leaders and managers.

INPO facilitates information exchange among member utilities by identifying and cataloging information on a wide range of activities that stations are doing especially well. The information on effective programs and practices is shared with members on request and through a number of other forums. This assistance fosters comparison and the exchange and emulation of successful methods among members.

i. **Assistance Visits**

Members may request assistance visits in specific areas of nuclear operations in which INPO personnel have experience or expertise. INPO personnel and industry peers normally conducts such visits. For example, if a member requests assistance in some specific aspect of maintenance, INPO will include a peer from another plant that handles that aspect of maintenance particularly well. INPO provides written reports that detail the results of the visits to the requesting utility. In most cases, the assistance visit includes actual methods and plans for improving performance as part of the assistance visit.

In 2009, INPO provided 144 assistance visits with 110 industry peers. Key areas of assistance provided included operational focus, maintenance and work management, engineering programs, chemistry, radiological protection, human performance, and industrial safety. Additional areas of assistance conducted in 2009 involved supplier participants, with a focus on supplemental personnel and fuel performance. In addition to assistance visits to stations for specific functional areas during 2009, senior representatives made 140 visits to their assigned stations to interact with station management and to monitor for early signs of performance decline. Senior representative-led INPO teams made 16 assistance visits at stations designated as special focus.

Effectiveness reviews performed by INPO approximately 6 months after assistance visits show that assistance visits are highly valued by station management and are contributing to improved performance.

ii. Development of Documents and Products

Several categories of documents and other products are designed and developed to help member utilities and participants achieve excellence in the operation, maintenance, training, and support of nuclear plants. Key categories of INPO documents and products are as follows:

- Principles documents address professionalism, management and leadership development, human performance, and other cross-functional topics important to achieving sustained operational excellence. INPO prepares these documents with substantial involvement of industry executives and managers. The principles extracted from the documents are used extensively in evaluation and assistance activities.

  The first of the principles documents entitled, *Principles for Enhancing Professionalism of Nuclear Personnel,* which addresses human resource management areas focused on developing nuclear professionals, including personnel selection, training and qualification, and career development. Two supplemental documents—*Management and Leadership Development* and *Excellence in Human Performance*—build on the original document. Utility executives use *Management and Leadership Development* to assist in the identification, development, assessment, and selection of future senior managers. *Excellence in Human Performance* provides practical suggestions for enhancements in the workplace that promote excellent human performance.

  In 1999, INPO distributed *Principles for Effective Self-Assessment and Corrective Action Programs*. This document emphasizes the importance of establishing a self-critical station culture and identifying the key elements of effective self-assessment and corrective action programs.

- Guideline documents establish the bases for sound programs in selected areas of plant operation, maintenance, and training, as well as cross-functional areas of direct importance to the operation and support of nuclear stations. Guidelines assist members in meeting the objectives used in evaluations and accreditation. The guidelines are recommendations based on generally accepted industry methods. They are not directives, but are intended to help utilities maintain high standards. Although member utilities do not have to follow each specific method described they are expected to strive to meet the intent of INPO guidelines.

- INPO provides good practices, work process descriptions, Nuclear Exchange documents, and other documents to assist members. Typically, these documents are developed from programs of member utilities and INPO's collective experience. INPO synthesizes the information into a document by the INPO staff, with industry input and review. In general, the documents define one method of meeting INPO performance objectives in specific areas, although other programs or methods may be as good or better. Utilities are encouraged to use these documents in developing or improving programs applicable to their plants. These documents can be used in whole or in part, as furnished, or modified to meet the specific needs of the plant involved.

INPO produces various other documents, such as analysis reports and special studies, as needed. Other assistance products include lesson plan materials, computer-based and interactive video materials, videotapes, and examination banks. The National Academy for Nuclear Training magazine, *The Nuclear Professional,* published quarterly, features how plant workers have solved problems and made improvements that enhanced safety.

iii. Workshops and Meetings

INPO sponsors workshops and working meetings for specific groups of managers on specific technical issues as forums for information exchange. This exchange provides an opportunity for INPO and industry personnel to discuss challenges, performance issues, and areas of interest. It also allows individuals from members and participants to meet and exchange information with their counterparts. In 2006, nearly 1,200 industry personnel participated in more than 70 meetings and workshops.

8. Key Initiatives 2010 – 2014

The nuclear industry continues to change and move at a demanding pace—new technologies, new people, and plans for new plants are adding even more challenges to the mix. The future will bring with it new demands for INPO and its members.

Cross-functional INPO teams began developing a strategic plan in mid-2008, building on the success and lessons learned from the previous plan. This was done by taking into account the needs of stakeholders and focusing on key areas in which INPO wants to have significant impact in the coming years.

The plan centers around four strategic focus areas:

SFA1: Increase accountability—both at INPO and in the industry—for full and timely resolution of adverse trends and issues.

SFA2: Advance industry performance in the areas of management, leadership, safety culture, recovery, and sustainability.

SFA3: Identify, develop, acculturate, and sustain a highly capable, professional, and knowledgeable workforce to lead and support nuclear organizations effectively.

SFA4: Advance nuclear safety worldwide using a network of partnerships that leverage our standards, methods, and global best practices to improve safe operations.

The 5-year business plan is built around high-priority organizational themes, critical for accomplishing INPO's vision. They are cross-functional, transcending cornerstone, division, and department boundaries. The plan is not a checklist of activities or projects that INPO does, but a plan that describes the outcomes INPO intends to produce or influence.

The industry continuously provides feedback to INPO on issues that affect station operation. Many INPO initiatives are based on industry trends and important focus areas. One initiative that is underway is described below.

a. New Plant Design and Construction

For many years, no new nuclear plants have been built in the U.S. However, as a result of the need for additional power, concerns over the environmental effects of carbon-based fuels, the streamlined licensing process, and financial incentives provided by the 2005 Energy Policy Act, U.S. utilities are once again planning new plant construction. To support this effort, INPO formed a new plant deployment group in 2006 to engage with the nuclear industry and plan for INPO's involvement though application of its cornerstone programs.

In 2006, INPO updated a report entitled, *Operating Experience to Apply to Advanced Light Water Reactors,* which includes lessons learned from significant events. The update report includes experience from operations and maintenance activities that should be addressed in the design of new plants. INPO participant plant designers and utility groups are using this document in their review of the new designs.

INPO also engaged utilities planning to submit license applications in a series of benchmarking trips in 2006 and 2007 to international utilities and plant designers in France and Japan, an aircraft company, and a coal plant with advanced control systems. These trips provided an opportunity to learn more about new technologies that have evolved since the last period of nuclear plant construction, most notably in plant standardization, computerized man-machine interface, and modular construction. INPO is promulgating a report to its members that features the information gathered from these trips.

To support plans for training the new plant workforce, INPO prepared a report entitled *Initial Accreditation of Training Programs for New Reactors,* which provides a process for achieving accreditation of training programs before their implementation. In addition, INPO will be reviewing the guidelines of the National Academy for Nuclear Training and several technical process description documents to make any necessary adjustments for the new plant environment.

9. Relationship with World Association of Nuclear Operators

U.S. nuclear utilities are represented in WANO through INPO. As such, INPO coordinates the U.S. nuclear utilities' activities in WANO. INPO also provides operational support and facilities for the WANO-Atlanta Centre, one of the four WANO global regional centers. The WANO-Atlanta Centre Governing Board usually appoints an INPO executive to serve as the Atlanta Centre director.

WANO-Atlanta Centre contracts with INPO to provide resources in terms of seconded staff to support the Centre's day-to-day operation. WANO-Atlanta Centre also contracts with INPO to provide administrative support services, such as payroll, computer support, and employee benefit administration.

WANO-Atlanta Centre activities and programs include the following:

- Peer reviews are conducted at the request of INPO members by WANO teams of U.S. and international peer reviewers who identify strengths and areas for improvement associated with nuclear safety and reliability. When conducted at a U.S. INPO member plant, a WANO peer review is performed in lieu of an INPO plant evaluation.
- WANO exchange of operating experience information provides detailed descriptions of events and lessons learned to member utilities worldwide.
- Performance indicator data are collected, trended, and disseminated to facilitate goal setting and performance trending and to encourage emulation of the best industry performance.
- Technical support missions are conducted to allow direct sharing of plant operating experience and ideas for improvement.
- Professional and technical development courses, seminars, and workshops are designed for enhancing staff development and sharing operating experience.

WANO-Atlanta Centre provides management and support services for the conduct of INPO's International Participant Program. This program facilitates the direct exchange of information and experience through INPO access to the secure member Web site, seminars, workshops, INPO documents, and exchange visits. International participants may chose to have liaison engineers located in the INPO offices for training and professional development to assist in the exchange of information. The international participants also provide INPO with advice on a wide range of nuclear-safety-related issues through membership on the International Participant Advisory Committee. The INPO International Participant Program is smaller in scope and complementary to the broader industry participation in WANO.

The U.S. industry and INPO receive a substantial benefit through their relationship with WANO and the international nuclear community. Many improvements have been implemented in the U.S. based on lessons learned from the more than 340 units that exist outside of the U.S. INPO works to remain fully aware of trends in the global nuclear industry and continues to strengthen relationships in this area.

10. Conclusion

The U.S. commercial nuclear industry has made substantial, sustained and quantifiable improvement in plant safety and performance during the three decades since the Three Mile Island event. The leaders who guided this industry over decades of challenge and change showed great insight when they recognized the need for an unprecedented form of industry self-regulation through peer review. The industry members acknowledged that nuclear energy would remain a viable form of electric power generation only if it could ensure the highest levels of nuclear safety and reliability (i.e., the achievement of excellence) in nuclear power plants. The industry responded to this challenge by creating an independent oversight process of the highest integrity and requiring of itself an uncompromising commitment to the standards and ethical principles that are essential to success.

This insight and commitment to integrity has provided the foundation for a unique, sustained partnership between INPO and its members. INPO is pleased to serve as an essential element of an industry that has raised its standards and improved its performance in nearly every aspect of plant operation. INPO does not take credit for this success but takes pride in its contribution to it.

INPO also recognizes that the pursuit of excellence is a continuing journey, not a destination. The U.S. nuclear industry, as it evolves and advances, will continue to encounter situations that challenge both people and equipment in a business environment that is competitive, complex, and increasingly global in character.

These challenges, while demanding, are not insurmountable. The U.S. commercial nuclear industry, in partnership with INPO, will continue the tradition of both sharing insight and acting with integrity, and in so doing, will continue on the shared journey to ever-higher levels of excellence.

APPENDIX A
NRC STRATEGIC PLAN 2008 - 2013

The U.S. Nuclear Regulatory Commission (NRC) published the NUREG-1614, Volume 4, "Strategic Plan: Fiscal Years 2008–2013" in February of 2008. This Appendix summarizes the key points of this plan.

A Stable Regulator in a Dynamic Environment

The regulatory environment associated with the use of radioactive materials is changing. The expected receipt of applications to construct and operate new nuclear power plants and to dispose of spent nuclear fuel and high-level radioactive waste, are two of the major challenges potentially facing the NRC over the next several years.

To meet these challenges, the NRC must efficiently use its resources, update the agency's regulatory review and construction inspection guidelines, and provide adequate infrastructure to accommodate staff.

Even as the NRC works to address growth in the industry, the agency's mission and values remain unchanged. The NRC's priority continues to be ensuring the adequate protection of public, health, safety, and the environment, while promoting the common defense and security.

Safety and security remain the agency's core functions, and the goals and strategic outcomes of the Strategic Plan are based on these functions. This focus on safety and security ensures that the NRC remains a strong, independent, stable, and predictable regulator.

Over the strategic planning period, the Nation is likely to see the following occur:

- The NRC expects to receive additional applications from entities that want to build and operate new nuclear power plants. The NRC also expects to receive applications for new fuel cycle facilities, including a significant number of uranium recovery applications.

- The U.S. Department of Energy (DOE) may submit an application to construct a high-level radioactive waste repository at Yucca Mountain, NV.[8]

- Increasing quantities of spent nuclear fuel will be held in interim storage at reactor sites or transported to centralized interim storage sites awaiting permanent disposal.

- The NRC will continue to coordinate with a wide array of Federal, State, local, and Tribal authorities on issues related to license renewal, new reactor licensing, homeland security, emergency planning, and protection of the environment.

- The number of NRC Agreement States will increase, as will the number of medical, academic, and industrial entities using radioactive materials under the oversight of the Agreement States.

The NRC recognizes that these changes will create an even greater need for effective and open communication with public stakeholders about a variety of issues. These issues include the

---

[8] In March 2010, DOE filed a motion to withdraw its application from NRC review. Section 19.8 of this report discusses radioactive waste in more detail.

safety and security of existing and proposed nuclear power plants and other licensed facilities and materials, emergency preparedness, and the impact on public health and safety and the environment from medical, academic, and industrial uses of licensed materials.

The unfolding of these complex regulatory issues also will require much more sophisticated techniques for the flow of documents and information, a process called knowledge management. The agency is in the process of attracting additional staff. The NRC realizes that to retain these highly skilled and educated professionals, who are critical to the agency, the agency must provide them with the necessary resources to do their jobs effectively and a high degree of workplace satisfaction. The agency's comprehensive knowledge management approach is focused on ensuring that all staff members are highly trained in the technical disciplines relating to their duties, the regulatory processes that govern agency actions, and the regulatory principles inherent in making the agency a strong, independent, stable, and predictable regulator.

Being a stable and predictable regulator implies having effective and structured regulatory processes in place and ensuring that these processes are followed. The agency will develop new regulatory initiatives in accordance with these processes, which will be open to public review and comment. The NRC is committed to considering and being responsive to stakeholder input before implementing any new regulatory initiative.

Key External Factors

The NRC's ability to achieve its goals depends on a changing mix of industry operating experience, national priorities, market forces, and availability of resources. A process for managing change should continue to be refined and implemented to ensure that the NRC is ready to address changing priorities in a timely manner. The following section discusses significant external factors, none of which the NRC can control but all of which could affect the agency's ability to achieve its strategic goals.

Receipt of New Reactor Operating License Applications. A resurgence of interest in new nuclear power plants is leading to intense competition for qualified individuals to serve as technical staff for both the NRC and its licensees and as nuclear power plant operating personnel. Increasing turnover and competition for qualified staff, as well as the loss of expertise as older members of the workforce retire, will remain an NRC challenge for the next several years.

Significant Operating Incident (Domestic or International). A significant incident at a licensed nuclear facility could cause the NRC to reassess its safety and security requirements, which could change the agency's focus on some initiatives related to its goals until the situation stabilizes. Because NRC stakeholders (including the public) are highly sensitive to many issues regarding the use of radioactive materials, events of relatively minor safety or security significance could potentially require a response that consumes considerable agency resources.

Significant Terrorist Incident. A significant terrorist incident anywhere in the United States would heighten the NRC's oversight and response stance. Subsequent new or changed security requirements or other policy decisions might affect the NRC, its partners, and the industry it regulates. A significant terrorist incident at a nuclear facility or activity anywhere in the world that departs from the agency's current evaluation of threat parameters could impact the NRC priorities, as well as U.S. policy regarding export activities, the NRC's role in international security, and requirements for security at U.S. nuclear power plants and other licensee facilities.

<u>Emergency Preparedness and Incident Response</u>.   Emergency preparedness and incident response activities with Federal, State, local, and Tribal authorities continue to increase in scope and number.   This affects the agency's priorities and workloads.

<u>Timing of the DOE Application and Related Activities for the High-Level Waste Repository at Yucca Mountain</u>.   The licensing of the proposed repository for spent nuclear fuel represents a major effort for the NRC in terms of planning, review, analysis, and ultimate decision-making. DOE has indicated that it intends to submit a license application for a high-level waste repository by June 2008.   The timing of DOE actions will heavily influence the NRC's resource allocation decisions over the next several years.   Acceleration or delay in DOE activities may affect other programs that are directly associated with achieving the agency's goals.[9]

<u>Legislative Initiatives</u>.   Legislative initiatives under consideration by the Congress can have a major impact on the NRC.   For example, the Energy Policy Act of 2005 has greatly affected the agency's priorities and workload.   Increasing interest in diversified sources of energy and energy independence is leading to an expected increase in license applications for nuclear power plants.   The attendant increase in resources devoted to license review and analysis is affecting how the agency goes about achieving its goals for this planning period.

<u>Advanced Fuel Cycle Development</u>.   DOE proposed the Global Nuclear Energy Partnership (GNEP) as a means to recycle (reprocess) nuclear fuel using proliferation-resistant technologies to recover more energy and reduce waste[10].   The impacts on the NRC could include developing the licensing requirements for, and then licensing, commercial reprocessing facilities, advanced burner reactors, and associated storage and waste facilities.   The scope and schedule of NRC activities are uncertain.

---

[9]   In March 2010, DOE filed a motion to withdraw its application from NRC review.   Section 19.8 of this report discusses radioactive waste in more detail.

[10]   In 2009, DOE cancelled the domestic GNEP program, focused primarily on domestic commercial recycling, and re-focused the program on continuation of research and development on proliferation-resistant fuel cycles and waste management strategies.

APPENDIX B
NRC MAJOR MANAGEMENT CHALLENGES FOR THE FUTURE

By law, the Inspector General of each Federal agency (discussed in Article 8 in Part 2 of this report) is to describe what he or she considers to be the most serious management and performance challenges facing the agency and assess the agency's progress in addressing those challenges. Accordingly, the Inspector General of the U.S. Nuclear Regulatory Commission (NRC) prepared his annual assessment of the major management challenges confronting the agency. The latest report, published in October 2009, can be found on the NRC's public Web site.

In his assessment, the Inspector General defined serious management challenges as "mission-critical areas or programs that have the potential for a perennial weakness or vulnerability that, without substantial management attention, would seriously impact agency operations or strategic goals." The challenges identified represent critical areas or difficult tasks that warrant high-level management attention. In the 2009 report, the Inspector General identified the following seven management challenges to be the most serious as of October 6, 2009.

Challenge 1: Protection of nuclear material used for civilian purposes

This challenge, which concerns materials control and accounting, is outside the scope of this report and is therefore not discussed.

Challenge 2: Managing information to balance security with openness and accountability

NRC employees often generate and work on sensitive information that needs to be protected. Such information can be sensitive unclassified information and classified national security information that is contained in written documents and electronic databases. In addressing continuing terrorist activity worldwide, the NRC continually reexamines its information management policies and procedures. The NRC faces the challenge of balancing the need to protect sensitive information from inappropriate disclosure with the agency's goal of openness in its regulatory processes. In 2008, the NRC made various efforts to improve public access to information while protecting sensitive information, including security-related information, from inappropriate disclosure.

Challenge 3: Ability to modify regulatory processes to meet a changing environment and to include the licensing of new nuclear facilities.

The NRC faces the challenge of maintaining its core regulatory programs while adapting to changes in its regulatory environment. The NRC must address a growing interest in licensing and constructing new nuclear power plants to meet the Nation's increasing demands for energy production. As of June 2009, the NRC had received 18 combined operating license (COL) applications and expects to receive an additional five COL applications by the end of fiscal year 2011.

While responding to the emerging demands associated with licensing and regulating new reactors, the NRC must maintain focus and effectively carry out its current regulatory responsibilities, such as inspections of the current fleet of operating nuclear reactors and fuel cycle facilities.   The NRC intends to increase its safety focus on licensing and oversight activities through risk-informed and performance-based regulation.

Challenge 4:   Oversight of radiological waste

The NRC regulates spent nuclear fuel generated from commercial nuclear power reactors, referred to here as high-level radioactive waste.   The NRC faces significant issues involving the potential licensing of the proposed repository for storing high-level radioactive waste located in Yucca Mountain, NV[11].   Additional challenges in the high-level waste area include the interim storage of spent nuclear fuel, certification of storage and transportation casks, and the oversight of decommissioned reactors and other nuclear sites.

Additionally, the amount of low-level waste continues to grow; however, no new disposal facilities have been built since the 1980s and unresolved issues will multiply as once-operational disposal facilities shut down.

Challenge 5:   Implementation of information technology and information security measures

The NRC needs to continue upgrading and modernizing its information technology and security capabilities both for employees and for public access to the regulatory process.   Recognizing the need to modernize, the Office of Information Services established goals to improve the productivity, efficiency, and effectiveness of agency programs and operations and to enhance the use of information for all users inside and outside the agency.   The NRC must also ensure that system security controls are in place to protect the agency's information systems against misuse.

Challenge 6:   Administration of all aspects of financial management

NRC management is responsible for establishing and maintaining effective internal controls and financial management systems that meet the objectives of several statutes including the Federal Managers' Financial Integrity Act.   This Act mandates that the NRC establish controls that reasonably ensure that (1) obligations and costs comply with applicable law; (2) assets are safeguarded against waste, loss, unauthorized use, or misappropriation; and (3) revenues and expenditures are properly recorded and accounted for.   This Act encompasses program, operational, and administrative areas, as well as accounting and financial management.

In addition, the NRC's management of its expanded grant program must be conducted in accordance with Federal regulations, which includes ensuring that funds are distributed and used as intended.

---

[11]   In March 2010, DOE filed a motion to withdraw its application from NRC review.   Section 19.8 of this report discusses radioactive waste in more detail.

Challenge 7:   Managing human capital

The NRC's human capital needs are changing in response to the receipt of applications to construct and operate the next generation of nuclear reactors and to increase the number of fuel cycle facilities.   To effectively manage human capital as these changes progress, while continuing to accomplish the agency's mission, the NRC must continue to implement the following initiatives:

- timely personnel security adjudication
- space planning
- recruitment, training, and knowledge management
- optimal use of resources

APPENDIX C
REFERENCES

American National Standards Institute (ANSI)

American National Standards Institute, ANSI N18.7-1976, "Administrative Controls and Quality Assurance for the Operational Phase of Nuclear Power Plants," February 1976.

American Society of Mechanical Engineers (ASME)

American Society of Mechanical Engineers, "Boiler and Pressure Vessel Code," Section XI, "Inservice Inspection of Nuclear Power Plant Components."

— — — Code Case N-716, "Piping Classification and Examination Requirements."

— — — Code Case N-770, "Alternative Examination Requirements and Acceptance Standards for Class 1 PWR Piping and Vessel Nozzle Butt Welds Fabricated with UNS N06082 or UNS W86182 Weld Filler Material With or Without Application of Listed Mitigation Activities," January 26, 2009.

— — — ASME-RA-5a-2009, "Standard for Level 1/Large Early Release Frequency Probabilistic Risk Assessment for Nuclear Power Plant Applications," 2009.

Boiling Water Reactor Vessel Internals Program (BWRVIP)

Boiling Water Reactor Vessel Internals Program, BWRVIP-75-A, "BWR Vessel and Internals Project, Technical Basis for Revisions to Generic Letter 88-01 Inspection Schedules," October 2005.

— — — BWRVIP-194, "Methodologies for Demonstrating Steam Dryer Integrity for Power Uprate," December 18, 2008.

— — — BWRVIP Letter 2007-051, William A. Eaton (BWRVIP Chairman) to BWRVIP Executive Committee, "Request for Information on Dissimilar Metal Weld Examinations," January 23, 2007.

— — — BWRVIP Letter 2007-062, Robin Dyle/Randy Stark to All BWRVIP Committee Members, "Letter to NRC Regarding BWRVIP Actions in Response to Flaw Indications in Recirculation Inlet Piping Nozzle to Safe End Welds at Duane Arnold," February 28, 2007.

— — — BWRVIP Letter 2007-139, Rick Libra (BWRVIP Chairman) to BWRVIP Executive Committee, "Request for Review of Dissimilar Metal Weld Examination Information," May 24, 2007.

— — — BWRVIP Letter 2007-367, Rick Libra (BWRVIP Chairman) to BWRVIP Executive Committee, "Recommendations Regarding Dissimilar Metal Weld Examinations (Includes Needed Requirement per NEI 03-08)," December 4, 2007.

Code of Federal Regulations

Title 10 of the *Code of Federal Regulations* Part 2, "Rules of Practice for Domestic Licensing Proceedings and Issuance of Orders."

——— Part 20, "Standards for Protection Against Radiation."

——— Part 21, "Reporting of Defects and Noncompliance."

——— Part 26, "Fitness for Duty Programs."

——— Part 34, "Licenses for Industrial Radiography and Radiation Safety Requirements for Industrial Radiographic Operations."

——— Part 35, "Medical Use of Byproduct Material."

——— Part 39, "Licenses and Radiation Safety Requirements for Well Logging."

——— Part 40, "Domestic Licensing of Source Material."

——— Part 50, "Domestic Licensing of Production and Utilization Facilities."

——— Part 51, "Environmental Protection Regulations for Domestic Licensing and Related Regulatory Functions."

——— Part 52, "Licenses, Certifications, and Approvals for Nuclear Power Plants."

——— Part 54, "Requirements for Renewal of Operating Licenses for Nuclear Power Plants."

——— Part 55, "Operator's Licenses."

——— Part 61, "Licensing Requirements for Land Disposal of Radioactive Waste."

——— Part 70, "Domestic Licensing of Special Nuclear Material."

——— Part 71, "Packaging and Transportation of Radioactive Material."

——— Part 72, "Licensing Requirements for the Independent Storage of Spent Nuclear Fuel and High-Level Radioactive Waste, and Reactor-Related Greater than Class C Waste."

——— Part 73, "Physical Protection of Plants and Materials."

——— Part 100, "Reactor Site Criteria."

——— Part 110, "Export and import of Nuclear Equipment and Material."

——— Part 140, "Financial Protection Requirements and Indemnity Agreements."

Title 40 of the *Code of Federal Regulations* Part 190, "Environmental Radiation Protection Standards for Nuclear Power Operations."

Title 44 of the *Code of Federal Regulations* Part 350, "Review and Approval of State and Local Radiological Emergency Plans and Preparedness."

Electric Power Research Institute

Electric Power Research Institute, "Materials Reliability Program: Primary System Piping Butt Weld Inspection and Evaluation Guideline," MRP-139, July 14, 2005.

— — — Topical Report 112657, "Revised Risk-Informed Inservice Inspection Evaluation Procedure," Rev. B-A, December 1999.

Federal Emergency Management Agency (FEMA)

Federal Emergency Management Agency, FEMA-REP-1 (See U.S. Nuclear Regulatory Commission, NUREG-0654.)

— — — Federal Policy Statement on Potassium Iodide Prophylaxis, January 2002.

GE Hitachi Nuclear Energy

GE Hitachi Nuclear Energy, NEDC-33436P, "GEH Boiling Water Reactor Steam Dryer—Plant Based Load Evaluation," November 7, 2008.

— — — NEDC-33408P, "ESBWR Steam Dryer - Plant Based Load Evaluation Methodology," February 2008.

International Atomic Energy Agency

International Atomic Energy Agency, "Guidelines for the Integrated Regulatory Review Service (IRRS)," February 2008.

— — — "Guidelines for the Integrated Regulatory Review Service (IRRS)," Edition 2010.

— — — Safety Series No. NS-G2.10, "Periodic Safety Review of Nuclear Power Plants, Vienna, August 2003.

— — — Safety Series No. 75-INSAG-4, "Safety Culture," Vienna, February 1991.

— — — Safety Series No. 115, "International Basic Safety Standards for Protection against Ionizing Radiation and for the Safety of Radiation Sources," Vienna, February 1996.

— — — TECDOC-953, "Method for the Development of Emergency Response Preparedness for Nuclear or Radiological Accidents," Vienna, 1997.

— — — TECDOC-955, "Generic Assessment Procedures for Determining Protective Actions During a Reactor Accident," Vienna, 1997.

International Commission on Radiological Protection (ICRP)

International Commission on Radiological Protection, ICRP Publication 26, "Recommendations of the International Commission on Radiological Protection," Oxford, Pergamon Press, January 1977.

— — — ICRP Publication 30, "Limits of Intakes of Radionuclides by Workers," eight volumes, Oxford, Pergamon Press, 1978–1982.

— — — ICRP Publication 60, "The 1990 Recommendations of the International Commission on Radiological Protection, Oxford," Pergamon Press, November 1990.

— — — ICRP Publication 103, "The 2007 Recommendations of the International Commission on Radiological Protection," Elsevier, March 2007.

Institute of Nuclear Power Operations (INPO)

Institute of Nuclear Power Operations, "Excellence in Human Performance," September 1997.

— — — "Initial Accreditation of Training Programs for New Reactors"

— — — "Management and Leadership Development," November 1994.

— — — "Operating Experience to Apply to Advanced Light Water Reactors," 2006.

— — — "Principles for a Strong Nuclear Safety Culture," November 2004.

— — — "Principles for Effective Self-Assessment and Corrective Action Programs," December 1999.

— — — "Principles for Enhancing Professionalism of Nuclear Personnel," March 1, 1989.

— — — Significant Operating Experience Report (SOER 02-4), "Reactor Pressure Vessel Head Degradation at Davis-Besse Nuclear Power Station," 2002.

National Council on Radiation Protection and Measurements (NCRP)

National Council on Radiation Protection and Measurements, NCRP Report No. 91, "Recommendations on Limits for Exposure to Ionizing Radiation," June 1987.

National Fire Protection Association (NFPA)

National Fire Protection Association, NFPA 805, "Performance-Based Standard for Fire Protection for Light Water Reactor Electric Generating Plants," 2001 Edition.

Nuclear Energy Institute (NEI)

Nuclear Energy Institute, NEI 00-01, "Guidance for Post-Fire Safe-Shutdown Circuit Analysis," Rev. 2, May 2009.

— — — NEI 04-02, "Guidance for Implementing a Risk-Informed, Performance-Based Fire Protection Program under 10 CFR 50.48(c)," Rev. 2, April 2008.

— — — NEI 04-04, "Cyber Security Programs for Power Reactors," November 18, 2005.

— — — NEI 07-13, "Methodology for Performing Aircraft Impact Assessments for New Plant Designs," May 2009.

— — — NEI 09-10, "Guidelines for Effective Prevention and Management of System Gas Accumulation," Rev. 0, October 2009.

— — — NEI 99-01, "Methodology for Development of Emergency Action Levels," Rev. 4, January 2003.

— — — "Buried Piping Integrity Initiative," November 18, 2009.

Nuclear Management and Resource Council (NUMARC)

Nuclear Management and Resource Council, NUMARC/NESP-007, "Methodology for Development of Emergency Action Levels," Rev. 2, January 1992.

U.S. Congress

Administrative Procedure Act, 5 U.S.C. Subchapter II

Atomic Energy Act of 1954, as amended, 42 U.S.C. 2011 et seq.

Debt Collection Improvement Act of 1996, 5 U.S.C. 5514 et seq.

DOE Organization Act, 42 U.S.C. 7101

Energy Policy Act of 2005, 16 U.S.C. 797 note et seq.

Energy Reorganization Act of 1974, as amended, 42 U.S.C. 5801 et seq.

Federal Civil Penalties Inflation Adjustment Act of 1990, 28 U.S.C. 2461

Hobbs Act (See Administrative Orders Review Act)

Homeland Security Act of 2002, 6 U.S.C. 101

National Environmental Policy Act of 1969, as amended, 42 U.S.C. 4321 et seq.

Nuclear Non-Proliferation Act of 1978, 22 U.S.C. 3201 et seq.

Price-Anderson Act of 1957, 42 U.S.C. 2012 et seq.

Uranium Mill Tailings Radiation Control Act of 1978, 42 U.S.C. 6907 et seq.

U.S. Department of Energy (DOE)

U.S. Department of Energy, DOE/EM-0654, "United States of America Third National Report for the Joint Convention on the Safety of Spent Fuel Management and on the Safety of Radioactive Waste Management," DOE in cooperation with the NRC, U.S. Environmental Protection Agency, and U.S. Department of State, Washington, DC, Rev. 2, October 2008.

U.S. Environmental Protection Agency

U.S. Environmental Protection Agency, EPA-400-R-92-001, "Manual of Protective Action Guides and Protective Actions for Nuclear Incidents," May 1992.

U.S. Nuclear Regulatory Commission (NRC)

U.S. Nuclear Regulatory Commission, "Agreement Between the Government of the United States of America and the Government of Canada on Cooperation in Comprehensive Civil Emergency Planning and Management" and "Administrative Arrangement Between the United States Nuclear Regulatory Commission and the Atomic Energy Control Board of Canada for Cooperation and the Exchange of Information in Nuclear Regulatory Matters," June 21, 1989.

——— "Agreement for the Exchange of Information and Cooperation in Nuclear Safety Matters" and "Implementing Procedure for the Exchange of Technical Information and Cooperation in Nuclear Safety Matters Between the Nuclear Regulatory Commission of the United States of America and the Comision Nacional de Seguridad Nuclear y Salvaguardias of Mexico," October 6, 1989.

——— "Alternate Fracture Toughness Requirements for Protection against Pressurized Thermal Shock Events," *Federal Register*, Vol. 75, January 4, 2010, p. 13, (75 FR 13).

——— "Annual Report on the Effectiveness of Training in the Nuclear Industry."

——— "Availability and Adequacy of Design Bases Information at Nuclear Power Plants; Policy Statement," *Federal Register,* Vol. 57, August 10, 1992, p. 35455, (57 FR 35455).

——— "Commission Policy Statement on the Systematic Evaluation of Operating Nuclear Power Reactors," *Federal Register,* Vol. 49, November 1984, p. 45112 (49 FR 45112).

——— Draft Guidance DG-1176, "Guidance for the Assessment of Beyond-Design-Basis Aircraft Impacts," July 2009.

——— Draft Guidance DG-1240, "Condition Monitoring Program for Electric Cables Used in Nuclear Power Plants," June 2010.

— — — Draft License Renewal Interim Staff Guidance, LR-ISG-2009-01, "Staff Guidance Regarding Plant-Specific Aging Management Review and Aging Management Program for Neutron-Absorbing Material in Spent Fuel Pools," November 23, 2009.

— — — "Draft Safety Culture Policy Statement: Request for Public Comments," *Federal Register,* Vol. 74, November 6, 2009, p. 57525 (74 FR 57525)

— — — Enforcement Guidance Memoranda, EGM 07-004, "Enforcement Discretion For Post-Fire Manual Actions Used as Compensatory Measures for Fire Induced Circuit Failures," June 30, 2007.

— — — Enforcement Guidance Memoranda, EGM 09-002, "Enforcement Discretion for Fire Induced Circuit Faults," May 14, 2009.

— — — Enforcement Guidance Memoranda, EGM 09-008, "Dispositioning Violations of NRC Requirements for Work Hour Controls Before and Immediately After a Hurricane Emergency Declaration," September 24, 2009.

— — — "Final Policy Statement on Technical Specification Improvements for Nuclear Power Reactors," *Federal Register,* Vol. 58, July 22, 1993, p. 39132 (58 FR 39132)

— — — Generic Letter 1982-04, "Use of INPO See-in Program," March 1982.

— — — Generic Letter 1982-33, "Requirements for Emergency Response Capability," December 17, 1982.

— — — Generic Letter 1996-04, "Boraflex Degradation in Spent Fuel Pool Storage Racks," June 26, 1996.

— — — Generic Letter 1981-38, "Storage of Low Level Radioactive Wastes at Power Reactor Sites," November 10, 1981

— — — Generic Letter 2007-01, "Inaccessible or Underground Power Cable Failures That Disable Accident Mitigation Systems or Cause Plant Transients," February 7, 2007.

— — — Generic Letter 2008-01, "Managing Gas Accumulation in Emergency Core Cooling, Decay Heat Removal and Containment Spray Systems," January 11, 2008.

— — — "Groundwater Task Force Final Report," June 30, 2010.

— — — Information Notice 87-43, "Gaps in Neutron-Absorbing Material in High-Density Spent Fuel Storage Racks," September 8, 1987.

— — — Information Notice 93-70, "Degradation of Boraflex Neutron Absorber Coupons," September 10, 1993.

— — — Information Notice 95-38, "Degradation of Boraflex Neutron Absorber in Spent Fuel Pool Storage Racks," September 8, 1995.

— — — Information Notice 97-78, "Crediting of Operator Actions in Place of Automatic Actions and Modifications of Operator Actions, Including Response Times," October 23, 1997.

— — — Information Notice 2004-05, "Spent Fuel Pool Leakage to Onsite Groundwater," March 3, 2004.

— — — Information Notice 2005-26, "Results of Chemical Effects Head Loss Tests in a Simulated PWR Sump Pool Environment," September 16, 2005.   (ADAMS Accession No. ML052570220)

— — — Information Notice 2005-26, Supplement 1, "Additional Results of Chemical Effects Tests in a Simulated PWR Sump Pool Environment," January 20, 2006. (ADAMS Accession No. ML060170102)

— — — Information Notice 2006-13, "Ground-Water Contamination Due to Undetected Leakage of Radioactive Water," July 10, 2006.

— — — Information Notice 2008-16, "Summary of Fitness-for-Duty Program Performance Reports for Calendar Year 2007," September 2, 2008.

— — — Information Notice 2009-26, "Degradation of Neutron-Absorbing Materials in the Spent Fuel Pool," October 28, 2009.

— — — Information Notice 2009-27, "Revised International Nuclear and Radiological Event Scale User's Manual," November 2009.

— — — Inspection Manual Chapter 0305, "Operating Reactor Assessment Program," December 24, 2009.

— — — Inspection Manual Chapter 2501, "Construction Inspection Program: Early Site Permit (ESP)," October 3, 2007.

— — — Inspection Manual Chapter 2502, "Construction Inspection Program: Pre-Combined License (Pre-COL) Phase," October 7, 2007.

— — — Inspection Manual Chapter 2503, "Construction Inspection Program: Inspections of Inspections, Tests, Analyses, and Acceptance Criteria (ITAAC)," October 3, 2007.

— — — Inspection Manual Chapter 2504, "Construction Inspection Program—Inspection of Construction and Operational Programs," October 15, 2009.

— — — Inspection Manual Chapter 2507, "Construction Inspection Program: Vendor Inspections," April 27, 2010.

— — — Inspection Manual Chapter 2509, "Browns Ferry Unit 1 Restart Project Inspection Program," September 2003.

— — — Inspection Manual Chapter 2517, "Watts Bar Unit 2 Construction Inspection Program," February 2008.

— — — Inspection Procedure 41500, "Training and Qualification Effectiveness," June 13, 1995.

— — — Inspection Procedure 42001, "Emergency Operating Procedures," June 28, 1991.

— — — Inspection Procedure 42700, "Plant Procedures," November 15, 1995.

— — — Inspection Procedure 71111.11, "Licensed Operator Requalification Program," January 5, 2006.

— — — Inspection Procedure 71152, "Problem Identification and Resolution," February 26, 2010.

— — — Inspection Procedure 71153, "Followup of Events and Notices of Enforcement Discretion," June 10, 2006.

— — — Inspection Procedure 95002, "Inspection for One Degraded Cornerstone or any Three White Inputs in a Strategic Performance Area", June 22, 2006.

— — — Inspection Procedure 95003, "Supplemental Inspection for Repetitive Degraded Cornerstones, Multiple Degraded Cornerstones, Multiple Yellow Inputs or One Red Input," October 2006 and November 9, 2009.

— — — "Inspector General's Assessment of the Most Serious Management and Performance Challenges Facing NRC," OIG-09-A-21, September 30, 2009.

— — — Interim Staff Guidance, "Digital Instrumentation and Controls DI&C-ISG-05 Task Working Group #5 Highly-Integrated Control Rooms—Human Factors Issues (HICR—HF) ISG," Rev. 1, November 3, 2008.

— — — "Open Government Plan," Rev. 1.1, June 7, 2010.

— — — "NRC Staff Review Guidance Regarding Generic Letter 2004-02 Closure in the Area of Plant-Specific Chemical Effect Evaluations," March 2008.

— — — NUREG-0090, Vol. 31, "Report to Congress on Abnormal Occurrences - Fiscal Year 2008," May 2009.

— — — NUREG-0396, "Planning Basis for the Development of State and Local Government Radiological Emergency Response Plans in Support of Light-Water Nuclear Power Plants, EPA-520/1-78/016," December 1978.

— — — NUREG-0654, "Criteria for Preparation and Evaluation of Radiological Emergency Response Plans and Preparedness in Support of Nuclear Power Plants," FEMA-REP-1, Rev. 1, November 1980.

— — — NUREG-0654, Supplement 3, "Criteria for Protective Action Recommendations for Severe Accidents (Draft Report for Interim Use and Comment)," Rev. 1, July 1996.

— — — NUREG-0700, "Human System Interface Design Review Guidelines," Rev. 2, May 2002.

— — — NUREG-0711, "Human Factors Engineering Program Review Model," Rev. 2, February 2004.

— — — NUREG-0713, "Occupational Radiation Exposure at Commercial Nuclear Power Reactors and Other Facilities," Vol. 30, January 2010.

— — — NUREG-0728, "NRC Incident Response Plan," Rev. 4, April 14, 2005.

— — — NUREG-0737, "Clarification of TMI Action Plan Requirements," November 1980.

— — — NUREG-0737, Supplement 1, "Requirements for Emergency Response Capability," January 1983.

— — — NUREG-0800, "Standard Review Plan for the Review of Safety Analysis Reports for Nuclear Power Plants," (formerly NUREG-75/087).

— — — NUREG-0847, Supplement 21, "Safety Evaluation Report Related to the Operation of Watts Bar Nuclear Plant, Unit 2," February 2009.

— — — NUREG-0899, "Guidelines for the Preparation of Emergency Operating Procedures," August 1982.

— — — NUREG-0933, "Resolution of Generic Safety Issues," August 2008.

— — — NUREG-1021, "Operator Licensing Examination Standards for Power Reactors," Rev. 9, Supplement 1, October 2007.

— — — NUREG-1022, "Event Reporting Guidelines, 10 CFR 50.72 and 50.73," Rev. 2, October 2000.

— — — NUREG-1055, "Improving Quality and the Assurance of Quality in the Design and Construction of Nuclear Power Plants, May 1984.

— — — NUREG-1150, "Severe Accident Risks: An Assessment for Five U.S. Nuclear Power Plants," December 1990.

— — — NUREG-1220, "Training Review Criteria and Procedures," Rev. 1, January 1993.

— — — NUREG-1350, Vol. 21, "NRC Information Digest 2009-2010," August 2009.

— — — NUREG-1430, "Standard Technical Specifications — Babcock and Wilcox Plants," June 2004.

— — — NUREG-1431, "Standard Technical Specifications — Westinghouse Plants," June 2004.

— — — NUREG-1432, "Standard Technical Specifications — Combustion Engineering Plants," June 2004.

— — — NUREG-1433, "Standard Technical Specifications — General Electric Plants (BWR/4)," June 2004.

— — — NUREG-1434, "Standard Technical Specifications — General Electric Plants (BWR/6)," June 2004.

— — — NUREG-1437, "Generic Environmental Impact Statement for License Renewal of Nuclear Power Plants," May 1996.

— — — NUREG-1465, "Accident Source Terms for Light-Water Nuclear Power Plants," February 1995.

— — — NUREG-1555, "Standard Review Plans for Environmental Reviews for Nuclear Power Plants," Supplement 1, "Operating License Renewal," March 2000.

— — — NUREG-1577, "Standard Review Plan on Power Reactor Licensee Financial Qualifications and Decommissioning Funding Assurance," Rev.1, February 1999.

— — — NUREG-1600, "General Statement of Policy and Procedures for NRC Enforcement Actions," November 28, 2008.

— — — NUREG-1614, Volume 4, "Strategic Plan: Fiscal Years 2008-20013," February 2008.

— — — NUREG-1650, "The United States Fourth National Report for the Convention on Nuclear Safety," Rev. 2, September 2007.

— — — NUREG-1764, "Guidance for the Review of Changes to Human Actions, Draft Report for Comment," Rev. 1, September 2007.

— — — NUREG-1791, "Guidance for Assessing Exemption Requests from the Nuclear Power Plant Licensed Operator Staffing Requirements Specified in 10 CFR 50.54(m)," July 2005.

— — — NUREG-1800, "Standard Review Plan for Review of License Renewal Applications for Nuclear Power Plants," Rev. 1, September 2005.

— — — NUREG-1801, "Generic Aging Lessons Learned (GALL) Report," Rev. 1, September 2005.

— — — NUREG-1852, "Demonstrating the Feasibility and Reliability of Operator Manual Actions in Response to Fire," October 2007.

— — — NUREG/BR-0150, Vol. 1, Rev. 4, "Response Technical Manual 96," March 1996.

— — — NUREG/CR-2850, "Dose Commitments Due to Radioactive Releases from Nuclear Power Plant Sites in 1992," Vol. 14, March 1996.

— — — NUREG/CR-6838, "Technical Basis for Assessing Exemptions from Nuclear Power Plant Licensed Operator Staffing Requirements 10 CFR 50.54(m)," February 2004.

— — — NUREG/CR-6847, "Cyber Security Self-Assessment Method for U.S. Nuclear Power Plants," October 2004.

— — — NUREG/CR-6850, "EPRI/NRC-RES Fire PRA Methodology for Nuclear Power Facilities," September 2005.

— — — NUREG/CR-6903, Vol. 1, "Human Event Repository and Analysis (HERA) System," July 2006.

— — — NUREG/CR-6903, Vol. 2, "Human Event Repository and Analysis (HERA) System," November 2007.

— — — NUREG/CR-6913, "Chemical Effects Head Loss Research in Support of Generic Safety Issue 191," December 2006.

— — — NUREG/CR-6914, "Integrated Chemical Effects Test Project," December 2006.

— — — NUREG/CR-7000, "Essential Elements of an Electric Cable Condition Monitoring Program," January 2010.

— — — OIG-09-A-18, "2009 NRC Safety Culture and Climate Survey," September 30, 2009.

— — — "OIG 2005 Survey of NRC's Safety Culture and Climate," OIG-06-A-08, February 10, 2006.

— — — Order EA-02-026, "Issuance of Order for Interim Safeguards and Security Compensatory Measures", February 25, 2002.

— — — Order EA-02-261, "Issuance of Order for Compensatory Measures Related To Access Authorization," January 7, 2003.

— — — Order EA-03-086, "Issuance of Order Requiring Compliance with Revised Design Basis Threat for Operating Power Reactors," April 29, 2003.

— — — "Palo Verde Nuclear Generating Station – NRC Integrated Inspection Report 05000528/2010002, 05000529/2010002, and 05000530/2010002," May 5, 2010.

— — — "Policy Statement on Use of PRA Methods in Nuclear Activities," *Federal Register,* Vol. 60, August 16, 1995, p. 42623, (60 FR 42623).

— — — Regulatory Guide 1.1, "Net Positive Suction Head for Emergency Core Cooling and Containment Heat Removal System Pumps," November 2, 1970.

— — — Regulatory Guide 1.3, "Assumptions Used for Evaluating the Potential Radiological Consequences of a Loss-of-Coolant Accident for Boiling-Water Reactors," Rev. 2, June 1974.

— — — Regulatory Guide 1.4, "Assumptions Used for Evaluating the Potential Radiological Consequences of a Loss-of-Coolant Accident for Pressurized-Water Reactors," Rev. 2, June 1974.

— — — Regulatory Guide 1.8, "Personnel Selection and Training," March 1971; Rev. 1, September 1975; Rev. 1-R, May 1977; and Regulatory Guide 1.8, "Qualification and Training of Personnel for Nuclear Power Plants," Rev. 2, April 1987; and Rev. 3, May 2000.

— — — Regulatory Guide 1.20, "Comprehensive Vibration Assessment Program for Reactor Internals during Preoperational and Initial Startup Testing," Rev. 3, March 2007.

— — — Regulatory Guide 1.21, "Measuring, Evaluating, and Reporting Radioactive Material in Liquid and Gaseous Effluents and Solid Waste," June 2009.

— — — Regulatory Guide 1.27, "Ultimate Heat Sink for Nuclear Power Plants," Rev. 2, January 1976.

— — — Regulatory Guide RG 1.28, "Quality Assurance Program Requirements (Design and Construction)," August 1985.

— — — Regulatory Guide 1.59, "Design Basis Floods for Nuclear Power Plants," Rev. 2, August 1977.

— — — Regulatory Guide 1.33, "Quality Assurance Program Requirements (Operations)," Rev. 2, February 1978.

— — — Regulatory Guide 1.70, "Standard Format and Content of Safety Analysis Reports for Nuclear Power Plants," Rev. 3, November 1978.

— — — Regulatory Guide 1.82, "Water Sources for Long-Term Recirculation Cooling Following a Loss of Coolant Accident," Rev. 3, November 2003.

— — — Regulatory Guide 1.101, "Emergency Planning and Preparedness for Nuclear Power Plants," Rev. 4, July 2003.

— — — Regulatory Guide 1.102, "Flood Protection for Nuclear Power Plants," Rev. 1, September 1976.

— — — Regulatory Guide 1.145, "Atmospheric Dispersion Models for Potential Accident Consequence Assessments at Nuclear Power Plants," Rev. 1, November 1982 (reissued with corrected page 1.145-7, February 1983).

— — — Regulatory Guide 1.147, "Inservice Inspection Code Case Acceptability, ASME Section XI, Division 1," Rev. 15, October 2007.

— — — Regulatory Guide 1.152, "Criteria for Use of Computers in Safety Systems of Nuclear Power Plants."

— — — Regulatory Guide 1.165, "Identification and Characterization of Seismic Sources and Determination of Safe Shutdown Earthquake Ground Motion," March 1997.

— — — Regulatory Guide 1.174, "An Approach for Using Probabilistic Risk Assessment in Risk-Informed Decisions on Plant-Specific Changes to the Licensing Basis," Rev. 1, November 2002.

— — — Regulatory Guide 1.177, "An Approach for Plant-Specific, Risk-Informed Decisionmaking: Technical Specifications," August 1998.

— — — Regulatory Guide 1.178, "An Approach for Plant-Specific Risk-Informed Decisionmaking for Inservice Inspection of Piping," September 2003.

— — — Regulatory Guide 1.183, "Alternative Radiological Source Terms for Evaluating Design Basis Accidents at Nuclear Power Reactors," July 2000.

— — — Regulatory Guide 1.188, "Standard Format and Content of Applications to Renew Nuclear Power Plant Operating Licenses," Rev. 1, September 2005.

— — — Regulatory Guide 1.189, "Fire Protection for Nuclear Power Plants," Rev. 2, November 2009.

— — — Regulatory Guide 1.200, "An Approach for Determining the Technical Adequacy of Probabilistic Risk Assessment Results for Risk-Informed Activities," Rev. 1, January 2007, and Rev. 2, March 2009.

— — — Regulatory Guide 1.201, "Guidelines for Categorizing Structures, Systems, and Components in Nuclear Power Plants According to Their Safety Significance," Rev. 1, May 2006.

— — — Regulatory Guide 1.205, "Risk-Informed, Performance-Based Fire Protection for Existing Light-Water Nuclear Power Plants," Rev. 0, May 2006, and Rev.1, December 2009.

— — — Regulatory Guide 1.206, "Combined Operating Licenses for Nuclear Power Plants," June 2007.

— — — Regulatory Guide 1.208, "A Performance-Based Approach to Define the Site-Specific Earthquake Ground Motion," March 2007.

— — — Regulatory Guide 4.1, "Radiological Environmental Monitoring for Nuclear Power Plants," June 2009.

— — — Regulatory Guide 4.2, "Preparation of Environmental Reports for Nuclear Power Stations," Rev. 2, July 1976.

— — — Regulatory Guide RG 4.21, "Minimization of Contamination and Radioactive Waste Generation: Life-Cycle Planning," June 2008.

— — — Regulatory Guide 5.71, "Cyber Security Programs for Nuclear Facilities," January 2010.

— — — Regulatory Guide 5.73, "Fatigue Management for Nuclear Power Plant Personnel," March 2009.

— — — Regulatory Guide 8.8, "Information Relevant to Ensuring That Occupational Radiation Exposures at Nuclear Power Stations Will Be As Low As Is Reasonably Achievable," Rev. 3, June 1978.

— — — Regulatory Issue Summary 2002-01, "Changes to NRC Participation in the International Nuclear Event Scale," January 2002.

— — — Regulatory Issue Summary 2006-10, "Regulatory Expectations with Appendix R Paragraph III.G.2 Operator Manual Actions," June 30, 2006.

— — — Regulatory Issue Summary 2007-06, "Regulatory Guide 1.200 Implementation," March 22, 2007.

— — — Regulatory Issue Summary 2008-25, "Regulatory Approach For Primary Water Stress Corrosion Cracking of Dissimilar Metal Butt Welds in Pressurized Water reactor Primary Coolant System Piping," October 22, 2008.

— — — RS-001, "Review Standard for Extended Power Uprates," December 2003.

— — — RS-002, "Processing Applications for Early Site Permits," May 3, 2004.

— — — Staff Requirements Memorandum on SECY-07-0096, "Possible Reactivation of Construction and Licensing Activities for the Watts Bar Nuclear Plant Unit 2," July 25, 2007.

— — —"Safety Goals for the Operation of Nuclear Power Plants; Policy Statement; Republication," *Federal Register*, Vol. 51, August 21, 1986, p. 30028, (51 FR 30028).

— — — SECY-07-0096, "Possible Reactivation of Construction and Licensing Activities for the Watts Bar Nuclear Plant Unit 2," June 7, 2007.

— — — SECY-08-0197, "Options to Revise Radiation Protection Regulations and Guidance with Respect to the 2007 Recommendations of the International Commission on Radiological Protection," December 18, 2008.

— — — SECY-09-0174, "Staff Progress in Evaluation of Buried Piping at Nuclear Reactor Facilities," December 2, 2009.

— — — SECY-09-0068, "Report of the Task Force on Internal Safety Culture," April 27, 2009.

— — — SECY-09-0143, "Status of the Accident Sequence Precursor Program and the Standardized Plant Analysis Risk Models," September 29, 2009.

— — — SECY-09-0159 "Annual Update of the Risk-Informed and Performance-Based Plan," October 27, 2009.

— — — SECY-10-0009, "Internal Safety Culture Update," January 26, 2010.

— — — SECY-10-0028, "FY 2009 Results of the Industry Trends Program for Operating Power Reactors and Status of Ongoing Development," March 16, 2010.

— — — SECY-10-0042, "Reactor Oversight Process Self-Assessment for Calendar Year 2009," April 7, 2010.

— — — SECY-93-087, "Policy, Technical, and Licensing Issues Pertaining to Evolutionary and Advanced Light-Water Reactor Designs," April 2, 1993.

— — —"Policy Statement on Severe Reactor Accidents Regarding Future Designs and Existing Plants," *Federal Register*, Vol. 50, August 8, 1985, p. 32138, (50 FR 32138).

— — — Temporary Instruction 2515/177, "Managing Gas Accumulation in Emergency Core Cooling, Decay Heat Removal, and Containment Spray Systems (NRC Generic Letter 2008-01)," June 9, 2009.

Westinghouse Electric Company

— — — WCAP-16125, "Justification for Risk-Informed Modifications to Selected Technical Specifications for Conditions Leading to Exigent Plant Shutdown," Rev. 1, December 2007.

— — — WCAP-16125, "Justification for Risk-Informed Modifications to Selected Technical Specifications for Conditions Leading to Exigent Plant Shutdown," Rev. 2, May 2009.

— — — WCAP-16308-NP, "Pressurized Water Reactor Owners Group 10 CFR 50.69 Pilot Program – Categorization Process – Wolf Creek Generating Station," Rev. 0, September 25, 2006.

— — — WCAP-16530-NP-A, "Evaluation of Post-Accident Chemical Effects in Containment Sump Fluids to Support GSI-191," March 2008.

— — — WCAP-16793-NP, "Evaluation of Long Term Cooling Considering Particulate, Fibrous and Chemical Debris in the Recirculating Fluid," April 2009.

Western European Nuclear Regulators' Association

Western European Nuclear Regulators' Association, "Pilot Study of Harmonization of Reactor Safety in WENRA Countries," March 2003.

Other Documents

Executive Order 12656 "Assignment of Emergency Preparedness Responsibilities," November 18, 1988.

Executive Order 12898, "Federal Actions To Address Environmental Justice in Minority and Low-Income Populations," *Federal Register*, Vol. 59, February 11, 1994, p. 7629, (59 FR 7629).

Homeland Security Presidential Directive 5, "Management of Domestic Incidents," March 4, 2003.

National Energy Policy, available at http://www.whitehouse.gov as of June 2007.

APPENDIX D
ABBREVIATIONS

ABWR	advanced boiling-water reactor
ADAMS	Agencywide Documents Access and Management System (NRC)
ALARA	as low as reasonably achievable
ANS	American Nuclear Society
ANSI	American National Standards Institute
AP	Advanced Passive
ASME	American Society of Mechanical Engineers
BRIIE	Baseline Risk Index for Initiating Events
BWR	boiling-water reactor
BWRVIP	Boiling-Water Reactor Vessel and Internals Project
CEO	chief executive officer
CFR	*Code of Federal Regulations*
CNS	Convention on Nuclear Safety
DHS	U.S. Department of Homeland Security
DOE	U.S. Department of Energy
EDG	emergency diesel generator
EGM	enforcement guidance memorandum
EPA	U.S. Environmental Protection Agency
EPIX	Equipment Performance Information Exchange database
EPR	evolutionary power reactor
EPRI	Electric Power Research Institute
EPU	extended power uprate
ERDA	U.S. Energy Research and Development Administration
ESBWR	economic simplified boiling-water reactor
FEMA	U.S. Federal Emergency Management Agency
FY	fiscal year
GE	General Electric
GL	generic letter
GNEP	Global Nuclear Energy Partnership
IAEA	International Atomic Energy Agency
ICRP	International Commission on Radiological Protection
IN	information notice
INPO	Institute of Nuclear Power Operations
IP	inspection procedure
IRRS	Integrated Regulatory Review Service
ISAP	Integrated Safety Assessment Program
ISG	interim staff guidance
ITAAC	inspection, test, analysis, and acceptance criterion/criteria

MWt	megawatt thermal
NANTeL	National Academy for Nuclear Training e-Learning
NCRP	National Council on Radiation Protection and Measurements
NEA	Nuclear Energy Agency
NEI	Nuclear Energy Institute
NEIL	Nuclear Electric Insurance Limited
NIMS	National Incident Management System
NRC	U.S. Nuclear Regulatory Commission
OMB	Office of Management and Budget
OSART	Operational Safety Assessment Review Team
POC	performance objectives and criteria
PRA	probabilistic risk assessment
PWR	pressurized-water reactor
RG	regulatory guide
RIS	regulatory issue summary
RISC	Risk-Informed Safety Class
RS	review standard
SAT	systems approach to training
SE	safety evaluation
SEE-IN	Significant Event Evaluation and Information Network
SEN	significant event notification
SEP	systematic evaluation program
SER	significant event report
SOER	significant operating experience report
SSC	structure, system, and component
Sv	sievert
TMI	Three Mile Island
TVA	Tennessee Valley Authority
US-APWR	U.S. Advanced Pressurized Water Reactor
US EPR	U.S. Evolutionary Power Reactor
WANO	World Association of Nuclear Operators
WENRA	Western European Nuclear Regulators' Association

APPENDIX E
ACKNOWLEDGMENTS

Contributors to this report include the following technical and regulatory experts from the U.S. Nuclear Regulatory Commission (NRC).

AbuEid, Boby
Alexion, Thomas
Anderson, Joseph
Arndt, Steven
Bagchi, Goutam
Bailey, Stewart
Beasley, Benjamin
Boatright, Aleem
Carpenter, Gene
Clarke, Deanna
Clayton, Brent
Collins, Jay
Cool, Don
Couret, Ivonne
Crockett, Steve
Crouch, Howard
Cubellis, Lou
Decker, David
Deegan, George
Dehn, Jeff
Demoss, Gary
Desaulniers, David
Dinsmore, Stephen
Dintz, Ira
Dudek, Michael
Dusaniwskyj, Michael
Fields, Leslie
Frahm, Ronald
Frumkin, Daniel
Galletti, Greg
Gallucci, Ray
Gartman, Michael
Gerke, Laura
Ghasemian, Shahram

Ghosh, Tina
Goodwin, Cameron
Gott, William
Graham, Throne
Gramm, Robert
Hackworth, Sandra
Hardies, Robert
Harrison, Donald
Hayden, Elizabeth
Hill, Britt
Hiser, Allan
Hopkins, Jon
Howe, Andrew
Huyck, Doug
Imboden, Andrew
Jarriel, Lisamarie
Jung, Ian
Kahler, Robert
Klein, Paul
Klementowicz, Stephen
Koltay, Peter
Kozal, Jason
Lain, Paul
Lobel, Rich
Lyon, Warren
Martin, Kamishan
Masciantonio, Armando
Meighan, Sean
Milano, Patrick
Miller, Barry
Mitchell, Matt
Muller, David
Munson, Clifford
Nourbakhsh, Hossein

Prescott, Paul
Rajone, Richard
Reed, Timothy
Reinert, Dustin
Rodriguez, Veronica
Rough, Richard
Sakai, Stacie
Salley, Mark
Samadar, Sujit
Scales, Kerby
Schaeffer, James
Schneider, Stewart
Schwartzman, Jennifer
Scott, Michael
Shoop, Undine
Sigmon, Rebecca
Simmons, Anneliese
Singal, Balwant
Snyder, Amy
Stewart, Sharon
Suttenberg, Jeremy
Tabatabai, Omid
Tailleart, Donald
Thompson, Jon
Titus, Brett
Virgilio, Rosetta
Williams, Barbara
Williams, Shawn
Wilson, George
Wilson, Jerry
Wong, Albert
Wong, Emma
Zalcman, Barry
Zimmerman, Jacob

Contributors to this report include the following experts from the Institute of Nuclear Power Operations (INPO).

David Farr

ANNEX 1
U.S. COMMERCIAL NUCLEAR POWER REACTORS

SOURCE: U.S. Nuclear Regulatory Commission NUREG-1350, Volume 21, "2009-2010 Information Digest," August 2009.

Plant Name and Operating Utility	Reactor Design Type	Licensed Power (MWt)	Operating Lifetime	
Arkansas Nuclear One 1 - Entergy Nuclear Operations, Inc.	PWR	2568	12/74	05/34
Arkansas Nuclear One 2 - Entergy Nuclear Operations, Inc.	PWR	3026	03/80	07/38
Beaver Valley 1 - FirstEnergy Nuclear Operating Company	PWR	2900	10/76	01/16
Beaver Valley 2 - FirstEnergy Nuclear Operating Company	PWR	2900	11/87	05/27
Braidwood 1 - Exelon Corp., Exelon Generation Co., LLC	PWR	3586.6	07/88	10/26
Braidwood 2 - Exelon Corp., Exelon Generation Co., LLC	PWR	3586.6	10/88	12/27
Browns Ferry 1 - Tennessee Valley Authority	BWR	3458	08/74	12/33
Browns Ferry 2 - Tennessee Valley Authority	BWR	3458	03/75	06/34
Browns Ferry 3 - Tennessee Valley Authority	BWR	3458	03/77	07/36
Brunswick 1 - Carolina Power & Light, Co., Progress Energy	BWR	2923	03/77	09/36
Brunswick 2 - Carolina Power & Light, Co., Progress Energy	BWR	2923	11/75	12/34
Byron 1 – Exelon Corp., Exelon Generation Co., LLC	PWR	3586.6	09/85	10/24
Byron 2 – Exelon Corp., Exelon Generation Co., LLC	PWR	3586.6	08/87	11/26
Callaway – AmerenUE, Union Electric Company	PWR	3565	12/84	10/24
Calvert Cliffs 1 - Constellation Energy	PWR	2700	05/75	07/34

Plant Name and Operating Utility	Reactor Design Type	Licensed Power (MWt)	Operating Lifetime	
Calvert Cliffs 2 - Constellation Energy	PWR	2700	04/77	08/36
Catawba 1 - Duke Energy Carolinas, LLC	PWR	3411	06/85	12/43
Catawba 2 - Duke Energy Carolinas, LLC	PWR	3411	08/86	12/43
Clinton - Exelon Corp., Exelon Generation Co., LLC	BWR	3473	11/87	09/26
Columbia Generating Station - Energy Northwest	BWR	3486	12/84	12/23
Comanche Peak 1- Luminant Generation Company, LLC	PWR	3612	08/90	02/30
Comanche Peak 2 - Luminant Generation Company, LLC	PWR	3458	08/93	02/33
Cooper - Nebraska Public Power District	BWR	2419	07/74	01/14
Crystal River 3 - Florida Power Corporation, Progress Energy	PWR	2609	03/77	12/16
Davis-Besse - FirstEnergy Nuclear Operating Co.	PWR	2817	07/78	04/17
Diablo Canyon 1 - Pacific Gas & Electric Co. D.C. Cook 1 - Indiana/Michigan Power Co.	PWR	3411	05/85	11/24
Diablo Canyon 2 - Pacific Gas & Electric Co. D.C. Cook 2 - Indiana/Michigan Power Co.	PWR	3411	03/86	08/25
Diablo Canyon 1 - Pacific Gas & Electric Co.Donald C. Cook 1 - Indiana/Michigan Power Co.	PWR	3304	08/75	10/34
Diablo Canyon 2 - Pacific Gas & Electric Co.Donald C. Cook 2 - Indiana/Michigan Power Co.	PWR	34683411	07/78	12/37
Dresden 2 - Exelon Corp., Exelon Generation Co., LLC	BWR	2957	06/70	12/29
Dresden 3 - Exelon Corp., Exelon Generation Co., LLC	BWR	2957	11/71	01/31
Duane Arnold - FPL Energy Duane Arnold, LLC, Florida Power and Light Co.	BWR	1912	02/75	02/14
Edwin I. Hatch 1 - Southern Nuclear Operating Co.	BWR	2804	12/75	08/34

Plant Name and Operating Utility	Reactor Design Type	Licensed Power (MWt)	Operating Lifetime	
Edwin I. Hatch 2 - Southern Nuclear Operating Co.	BWR	2804	09/79	06/38
Fermi 2 – The Detroit Edison Co.	BWR	3430	01/88	03/25
Fort Calhoun Station – Omaha Public Power District	PWR	1500	09/73	08/33
R.E. Ginna - Constellation Energy	PWR	1775	07/70	09/29
Grand Gulf 1 - Entergy Nuclear Operations, Inc.	BWR	3898	07/85	11/24
H.B. Robinson 2 - Carolina Power & Light Co.	PWR	2339	03/71	07/30
Hope Creek 1 - PSEG Nuclear, LLC	BWR	3840	12/86	04/26
Indian Point 2 - Entergy Nuclear Operations, Inc.	PWR	3216	08/74	09/13
Indian Point 3 - Entergy Nuclear Operations, Inc.	PWR	3216	08/76	12/15
James A. FitzPatrick - Entergy Nuclear Operations, Inc.	BWR	2536	07/75	10/34
Joseph M. Farley 1 - Southern Nuclear Operating Co.	PWR	2775	12/77	06/37
Joseph M. Farley 2 - Southern Nuclear Operating Co.	PWR	2775	07/81	03/41
Kewaunee Power Station - Dominion Energy Kewaune, Inc.	PWR	1772	06/74	12/13
La Salle County 1 - Exelon Corp., Exelon Generation Co., LLC	BWR	3489	01/84	04/22
La Salle County 2 - Exelon Corp., Exelon Generation Co., LLC	BWR	3489	10/84	12/23
Limerick 1-Exelon Corp., Exelon Generation Co., LLC	BWR	3458	02/86	10/24
Limerick 2- xelon Corp., Exelon Generation Co., LLC	BWR	3458	01/90	06/29
McGuire 1 - Duke Energy Power Company, LLC	PWR	3411	12/81	06/41
McGuire 2 - Duke Energy Power Company, LLC	PWR	3411	03/84	03/43
Millstone 2 – Dominion Nuclear Connecticut, Inc., Dominion Generation	PWR	2700	12/75	07/35

Plant Name and Operating Utility	Reactor Design Type	Licensed Power (MWth)	Operating Lifetime	
Millstone 3 - Dominion Nuclear Connecticut, Inc., Dominion Generation	PWR	3650	04/86	11/45
Monticello - Nuclear Management Co.	BWR	1775	06/71	09/30
Nine Mile Point 1 - Constellation Energy	BWR	1850	12/69	08/29
Nine Mile Point 2 - Constellation Energy	BWR	3467	03/88	10/46
North Anna 1 Virginia Electric & Power Co., Dominion Generation	PWR	2893	06/78	04/38
North Anna 2 - Virginia Electric & Power Co., Dominion Generation	PWR	2893	12/80	08/40
Oconee 1 - Duke Energy Power Company, LLC	PWR	2568	07/73	02/33
Oconee 2 - Duke Energy Power Company, LLC	PWR	2568	09/74	10/33
Oconee 3 - Duke Energy Power Company, LLC	PWR	2568	12/74	12/34
Oyster Creek - AmerGen Energy Co., LLC, Exelon Corp.	BWR	1930	12/69	04/29
Palisades - Entergy Nuclear Operations, Inc.	PWR	2565	12/71	03/31
Palo Verde 1 - Arizona Public Service Company	PWR	3990	01/86	06/25
Palo Verde 2 - Arizona Public Service Company	PWR	3990	09/86	04/26
Palo Verde 3 - Arizona Public Service Company	PWR	3990	01/88	11/27
Peach Bottom 2 Exelon Corp., Exelon Generation Co., LLC	BWR	3514	07/74	08/33
Peach Bottom 3 Exelon Corp., Exelon Generation Co., LLC	BWR	3514	12/74	07/34
Perry 1 - FirstEnergy Nuclear Operating Co.	BWR	3758	11/87	03/26
Pilgrim 1 - Entergy Nuclear Operations, Inc.	BWR	2028	12/72	06/12
Point Beach 1 - FLP Energy Point Beach, LLC, Florida Power and Light Co.	PWR	1540	12/70	10/30
Point Beach 2 - FLP Energy Point Beach, LLC, Florida Power and Light Co.	PWR	1540	10/72	03/33
Prairie Island 1 - Nuclear Management Co.	PWR	1650	12/73	08/13

Plant Name and Operating Utility	Reactor Design Type	Licensed Power (MWth)	Operating Lifetime	
Prairie Island 2 - Nuclear Management Co.	PWR	1650	12/74	10/14
Quad Cities 1 Exelon Corp., Exelon Generation Co., LLC	BWR	2957	02/73	12/32
Quad Cities 2 - Exelon Corp., Exelon Generation Co., LLC	BWR	2957	03/73	12/32
River Bend 1 - Entergy Nuclear Operations, Inc.	BWR	3091	06/86	08/25
Salem 1 - PSEG Nuclear, LLC	PWR	3459	06/77	08/16
Salem 2 - PSEG Nuclear, LLC	PWR	3459	10/81	04/20
San Onofre 2 - Southern California Edison Co.	PWR	3438	08/83	02/22
San Onofre 3 - Southern California Edison Co.	PWR	3438	04/84	11/22
Seabrook 1 - FPL Energy Seabrook, LLC	PWR	3648	08/90	03/30
Sequoyah 1 - Tennessee Valley Authority	PWR	3455	07/81	09/20
Sequoyah 2 - Tennessee Valley Authority	PWR	3455	06/82	09/21
Shearon Harris 1 - Carolina Power & Light Co.	PWR	2900	05/87	10/46
South Texas Project 1 - STP Nuclear Operating Co.	PWR	3853	08/88	08/27
South Texas Project 2 - STP Nuclear Operating Co.	PWR	3853	06/89	12/28
St. Lucie 1 - Florida Power & Light Co.	PWR	2700	12/76	03/36
St. Lucie 2 - Florida Power & Light Co.	PWR	2700	08/83	04/43
Surry 1 - Dominion Generation	PWR	2546	12/72	05/32
Surry 2 - Dominion Generation	PWR	2546	05/73	01/33
Susquehanna 1 - PPL Susquehanna, LLC	BWR	3952	06/83	07/22
Susquehanna 2 - PPL Susquehanna, LLC	BWR	3952	02/85	03/24
Three Mile Island 1 - AmerGen Energy Co., LLC	PWR	2568	09/74	04/14
Turkey Point 3 - Florida Power & Light Co.	PWR	2300	12/72	07/32

Plant Name and Operating Utility	Reactor Design Type	Licensed Power (MWth)	Operating Lifetime	
Turkey Point 4 - Florida Power & Light Co.	PWR	2300	09/73	04/33
V.C. Summer - South Carolina Electric & Gas Co.	PWR	2900	01/84	08/42
Vermont Yankee - Entergy Nuclear Operations, Inc.	BWR	1912	11/72	03/12
Vogtle 1 - Southern Nuclear Operating Co.	PWR	3625	06/87	01/47
Vogtle 2 - Southern Nuclear Operating Co.	PWR	3625	05/89	02/49
Waterford 3 - Entergy Nuclear Operations, Inc	PWR	3716	09/85	12/24
Watts Bar 1 - Tennessee Valley Authority	PWR	3459	05/96	11/35
Wolf Creek 1 - Wolf Creek Nuclear Operating Corp.	PWR	3565	09/85	03/45

ANNEX 2
U.S. NUCLEAR ELECTRIC INDUSTRY
PERFORMANCE INDICATOR GRAPHS

Unit Capability Factor
1-Year Median Values
December 2009

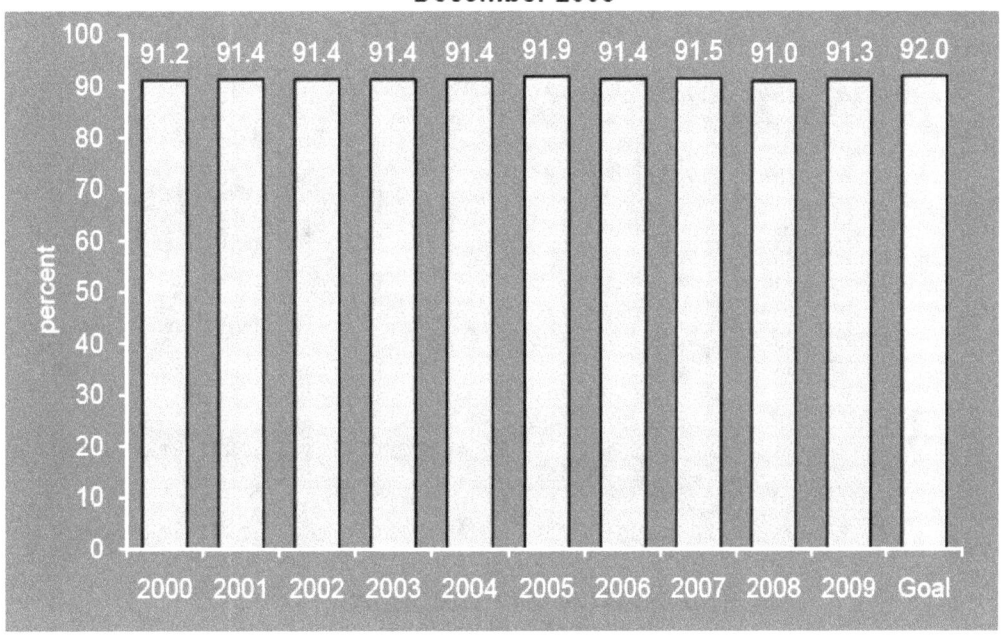

Forced Loss Rate
1-Year Median Values
December 2009

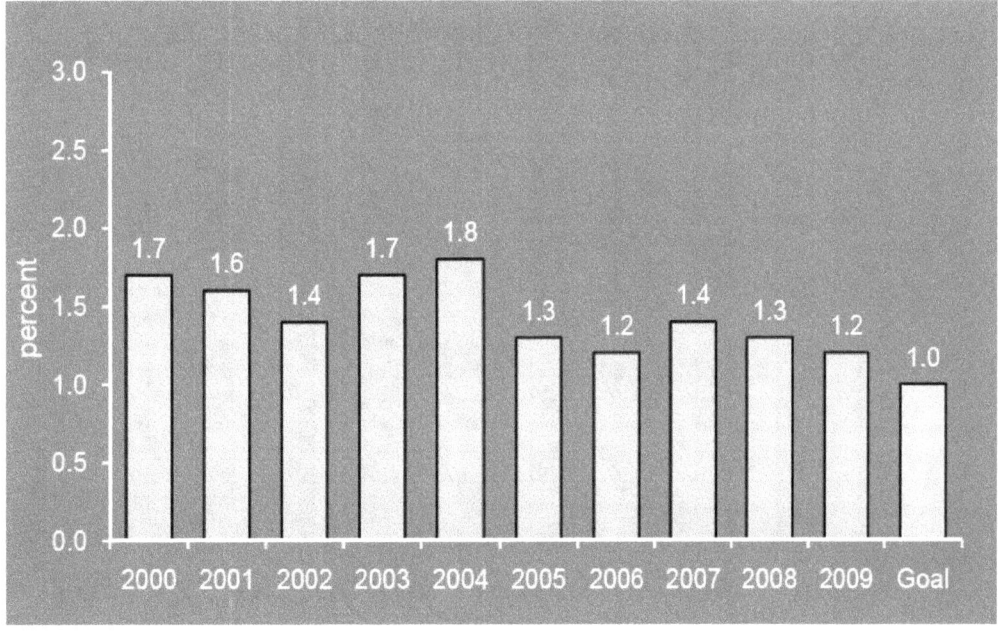

Unplanned Automatic Scrams
1-Year Median Values
December 2009

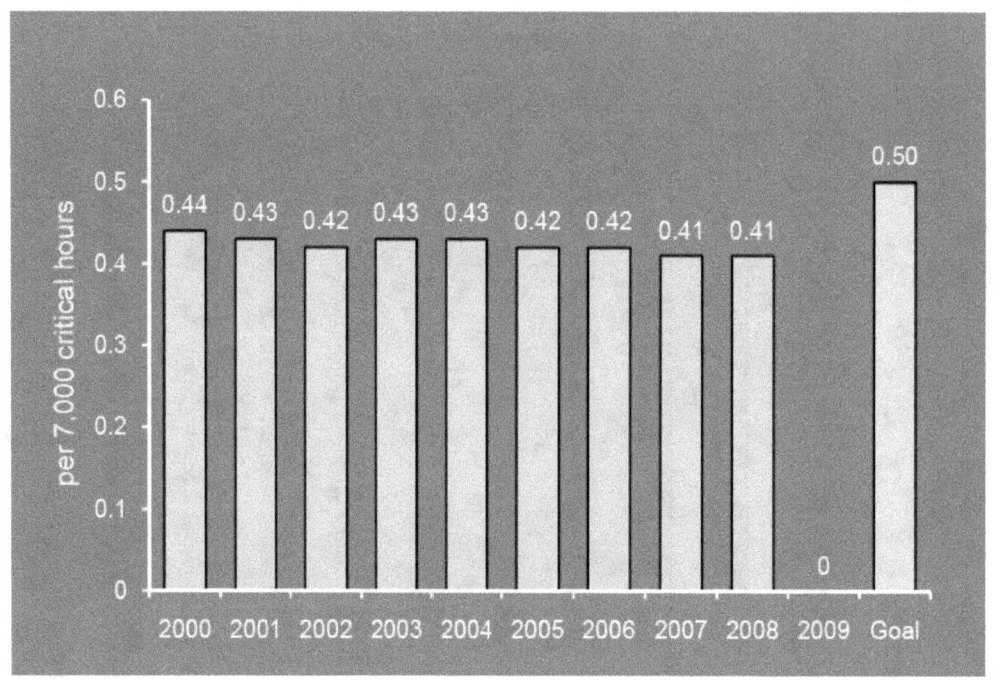

Safety System Performance
1-Year Median Values
December 2009

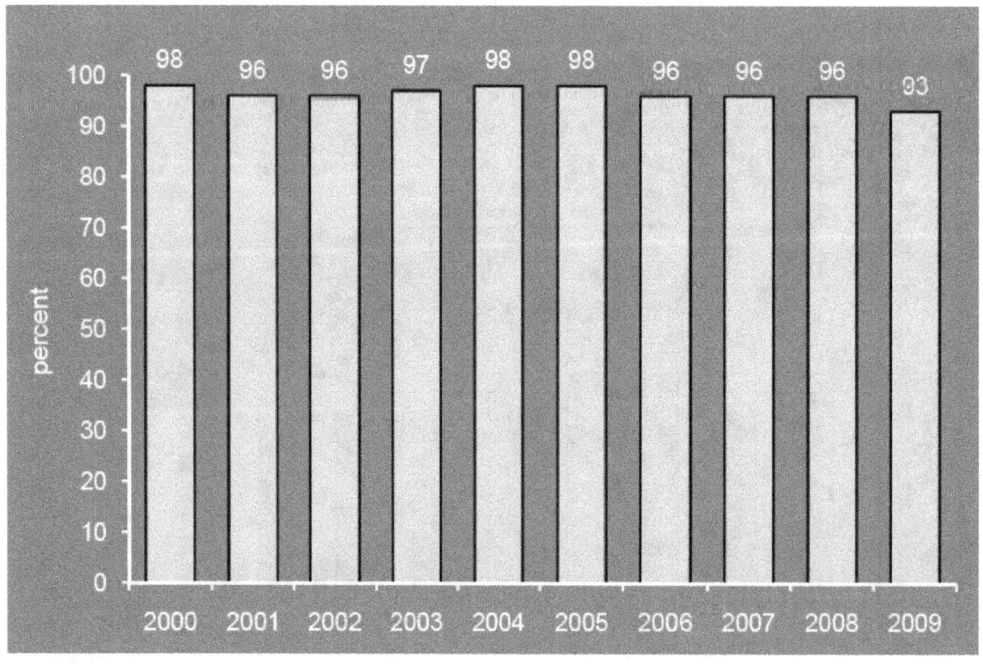

Fuel Reliability
1-Year Median Values
December 2009

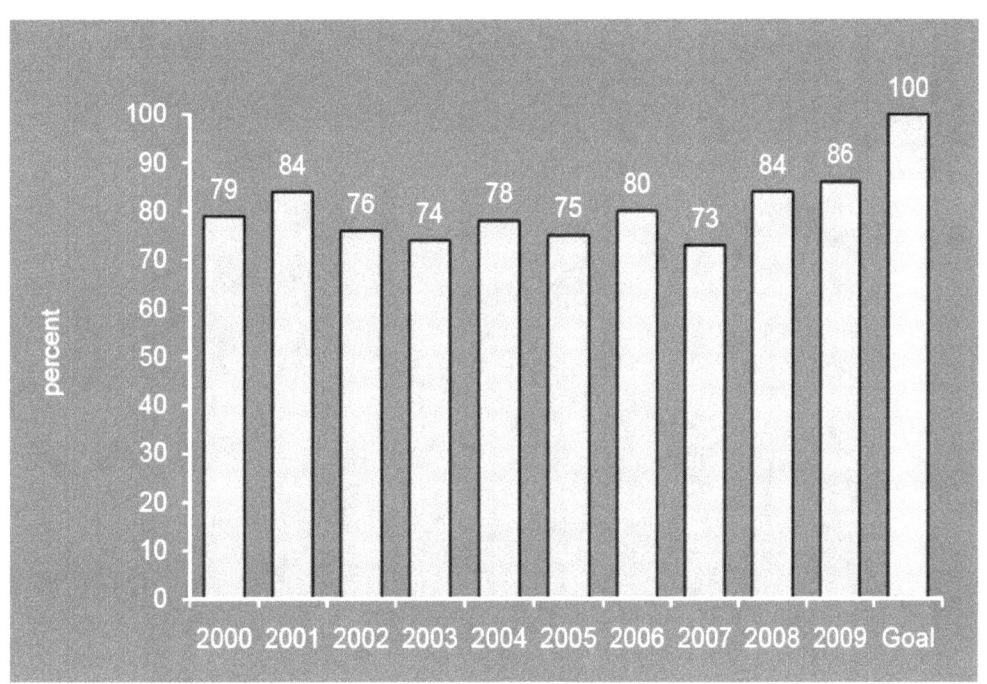

Collective Radiation Exposure (BWR)
1-Year Median Values
December 2009

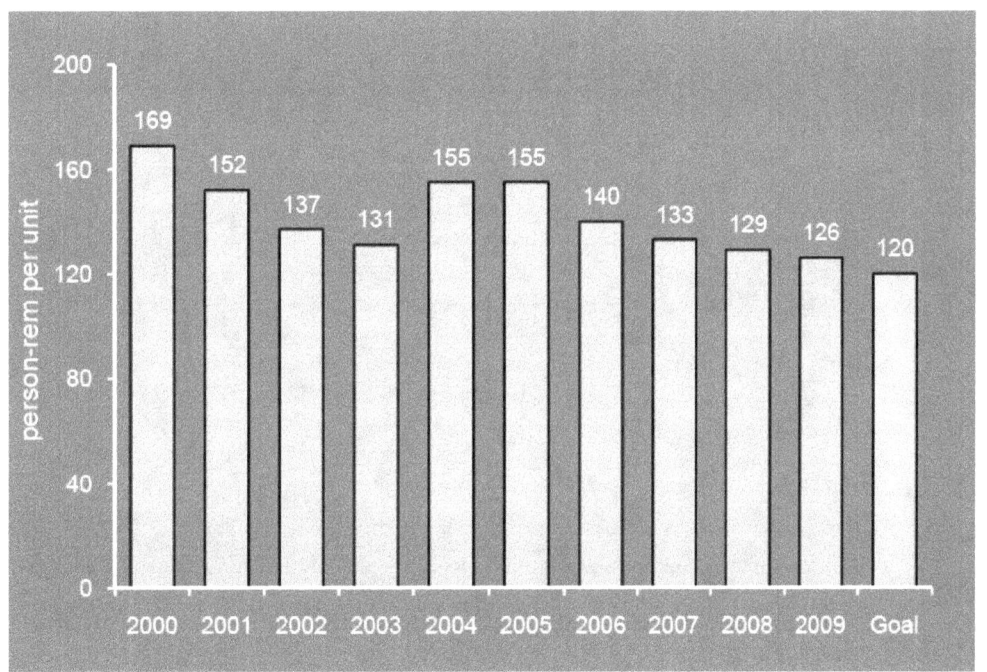

Collective Radiation Exposure (PWR)
1-Year Median Values
December 2009

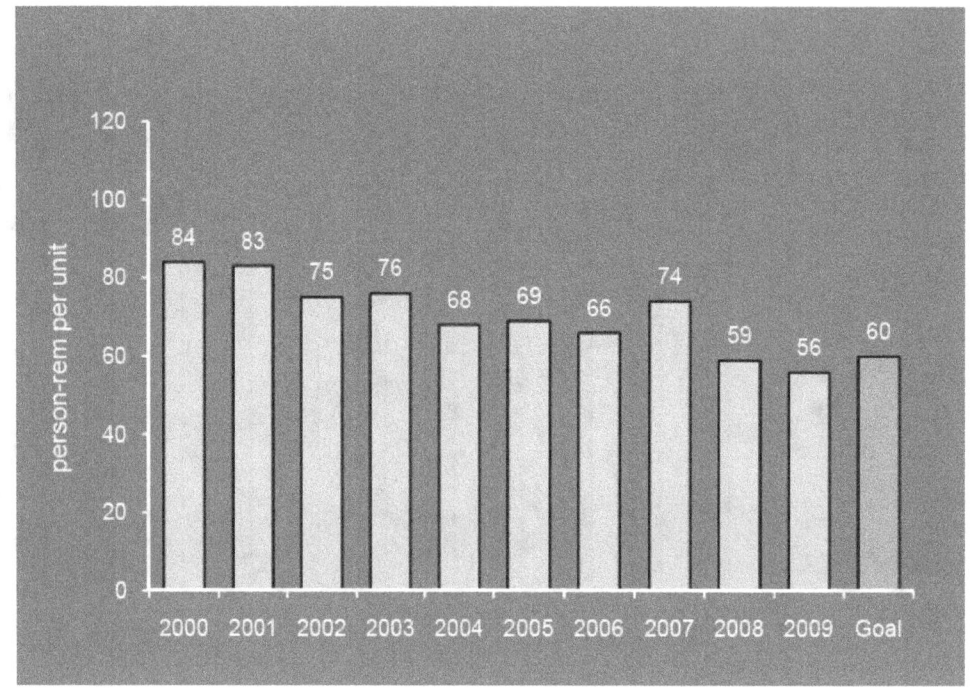

Total Industrial Safety Accident Rate
1-Year Median Values
December 2009

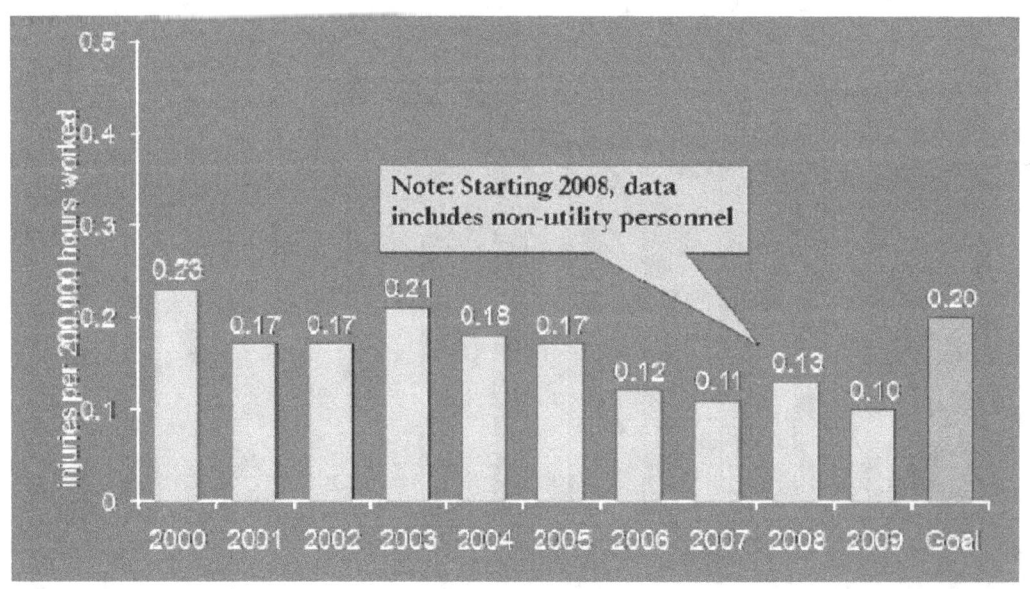